THE EXPLORER'S GENE

ALSO BY ALEX HUTCHINSON

Endure

Which Comes First, Cardio or Weights?

Big Ideas: 100 Modern Inventions That Have Transformed Our World

THE EXPLORER'S GENE

Why We Seek Big Challenges, New Flavors, and the Blank Spots on the Map

Alex Hutchinson

MARINER BOOKS

New York   Boston

THE EXPLORER'S GENE. Copyright © 2025 by Alex Hutchinson. All rights reserved. Printed in the United States of America. No part of this book may be used or reproduced in any manner whatsoever without written permission except in the case of brief quotations embodied in critical articles and reviews. For information, address HarperCollins Publishers, 195 Broadway, New York, NY 10007.

HarperCollins books may be purchased for educational, business, or sales promotional use. For information, please email the Special Markets Department at SPsales@harpercollins.com.

The Mariner flag design is a registered trademark of HarperCollins Publishers LLC.

FIRST EDITION

Designed by Chloe Foster
Cartography by Julie Witmer Custom Map Design

Library of Congress Cataloging-in-Publication Data has been applied for.

ISBN 978-0-06-326976-7

24 25 26 27 28 LBC 5 4 3 2 1

*For Ella and Natalie, who sparked
the questions in these pages.*

To say that we should not change wines is a heresy; the tongue becomes saturated, and after the third glass even the best bottle yields but an obtuse sensation.

—Jean Anthelme Brillat-Savarin, *Physiologie du goût*

CONTENTS

Introduction   1

PART I: WHY WE EXPLORE

1. The Great Human Expansion   17
2. Chasing Dopamine   36
3. The Free Energy Principle   53

PART II: HOW WE EXPLORE

4. Explore vs. Exploit   75
5. Unknown Unknowns   96
6. The Time Horizon   114
7. Mapping the World   130
8. The Landscape of Ideas   149

PART III: WHAT EXPLORING MEANS NOW

9. The Problem with Passive   169
10. Rediscovering Play   188

11.	The Effort Paradox	203
12.	The Future of Exploring	221
	Epilogue	241
	Acknowledgments	247
	Notes	249
	Index	279

INTRODUCTION

It's the single most iconic vista in all of Newfoundland, all the more prized because it's so hard to reach. By the time we clambered over the final set of boulders to get there, we'd been climbing for more than six hours, accompanied by clouds of voracious and seemingly waterproof blackflies that were undeterred by the steadily falling rain. We turned to look back at the route we'd traveled: the sinuous, glacier-carved fjord two thousand feet below us, the billion-year-old cliffs that hemmed it in, the jumble of rocks and rain forest that led steeply up to the plateau where we now stood. This view of Western Brook Pond is a staple of the island's glossy tourism campaigns; we've seen the pics, but on that particular day it was nothing but a blanket of mist.

We didn't have time to linger anyway. It was nearly noon by the time the boat had dropped us off at the head of the fjord, then climbing up the gulch had taken twice as long as we'd anticipated. We were barely halfway to the alpine pond where we'd hoped to camp that night. As the mist thickened, finding landmarks was becoming increasingly difficult. Muddy game trails carved by the area's ubiquitous moose and caribou led in every direction through the boggy grass, frequently disappearing into sinkholes filled by several days of nonstop rain. No matter how often we stopped to orient ourselves, we were turned around again within minutes.

I felt panic rising in me. We were already a day behind schedule, because the waters of the fjord had been too choppy for the boat on our scheduled departure day. That had forced us to burn a day of food while camped by the dock waiting for our ride, leaving us with just four days to complete the hike instead of the planned five. And while my wife, Lauren, and I were capable of hiking as long into the night as we needed to, we couldn't ask the same of our daughters, Ella and Natalie. They were just 8 and 6, respectively—and, aside from being exhausted, they were being driven bonkers by the flies, despite their full-body bug suits. But there were no exits from this hike. No roads traverse this part of Newfoundland. The boat was gone, and so was our cell signal. The only way out was onward. In that moment of maximal uncertainty, a puzzling thought nagged at me.

"You know," I said to Lauren, "this isn't bad planning or bad luck. It's *exactly* what we asked for."

We've been backpacking and canoeing with our kids since they were a few months old, and have put a lot of thought into the routes we choose. We want challenge and adventure, but also safety and pleasure and variety and natural beauty, titrated each year to the kids' steadily expanding capabilities. They had already canoed in Algonquin Park, hiked in the Rockies, backpacked on the Bruce Peninsula. So when we planned our 2022 trip, we were looking for something with a twist—something that would feel like a voyage of discovery as much for Lauren and me as for the kids.

We found the twist on Parks Canada's official website for Gros Morne National Park, on Newfoundland's sparsely inhabited western coast. The park is a major tourist attraction—but like most such parks, the visitors generally stick to a few easily accessible places. What Bill Bryson wrote about U.S. national parks tends to be true around the world: "98 percent of visitors arrive by car, and 98 percent of those venture no more than 400 yards from their metallic wombs." In contrast, the Long Range Traverse—"an unmarked and rugged backcountry route," according to the website—cuts across the park's interior. It covers twenty-two miles as the crow flies (or at least as the mapmaker, sitting in a comfortable room with a piece of string, calculates) between the head of an inland fjord called Western Brook Pond and the base of Gros Morne Mountain. The difficult terrain, impenetrable vegetation, and challenging navigation make it impossible to follow the shortest path, so most hikers cover at least 50 percent more distance than that. Only three groups, with a maximum of four people per group, are allowed to start each day. The day before you start, you have to attend a safety briefing and demonstrate your navigational skills. Rescues along the route are extremely challenging but all too frequent. "We are, therefore, encouraging visitors to opt for less risky adventures," the Parks Canada site warned.

That's the line that lured us in: rather than singing the praises of their beautiful hiking route, the park was begging people to stay away. We understood the risks, but we figured that our prior backcountry experience would enable us to do it safely, and that a hike billed as three to four nights was within the physical capacities our kids had already demonstrated. The idea of a hike with no actual trail, where we would have to use our own best judgment to pick the optimal route through a swath of untrammeled wilderness, was irresistible. And it was perfectly in keeping with the way Lauren and I had approached our vacations ever since we started dating, nearly two decades earlier.

When we met, I was living in Washington, D.C., and Lauren was living in South Bend, Indiana. We went on our first date just after Christmas in 2003, while we were both visiting our families in Toronto. I visited her for a weekend in Indiana a few months later; she spent Easter in Washington. By then we were already planning what was effectively our fourth date: a ten-day backpacking trip in Alberta. We'd been contemplating routes in Banff and Jasper national parks when my cousin told us about the Willmore Wilderness Park, an obscure protected area north of Jasper that's about 50 percent bigger than Yosemite. It has no official trails, no rangers, no facilities, and is hundreds of miles from the nearest airport. It's the same awe-inspiring mountains as the crowded parks farther south, just much harder to get to and to find your way around.

We saw other humans just once during our ten days in Willmore, and that trip remains the template against which I measure all others. We hiked long hours up imposing passes to get deep into the mountains—and once we were there, we could look around, pick an interesting-looking peak, and spend the afternoon scrambling to the top. I remember taking in the view from one of these peaks, looking around as far as we could see in the distance and finding no signs of humanity in any direction—except for a tiny green dot migrating down the pass below us, which we eventually realized was our tent, blown free from its moorings by a vicious alpine wind.

In the years since, on travels both with and without Lauren, I've always opted for the less obvious, less-well-trodden destinations. In Australia, we loved the Ningaloo Reef, hundreds of miles north of Perth on the barren west coast, far more than the Great Barrier Reef. In India to cover the Commonwealth Games as a journalist, I skipped a prepackaged day trip to the Taj Mahal in order to see if I could make it to the Red Fort from my aseptic hotel by wandering several miles on foot through the crooked alleyways of Old Delhi. On New Zealand's South Island, I read all about the Milford Track, famously dubbed "the finest walk in the world" by the *Spectator* way back in 1908—and then chose to hike the less heralded Routeburn Track instead.

When I began writing adventure travel journalism for the *New York Times* in the late 2000s, I noticed that, without intending to, I kept circling back to this theme. On a canoe trip deep in the Yukon wilderness, I bushwhacked up nondescript ridges precisely because they seemed so unlikely to have attracted anyone else's interest. "It was intoxicating," I wrote, "to pick a point in the distance and wonder: Has any human ever stood there?" Backpacking along the remote southern coast of Tasmania—a route whose impenetrable landscape and miserable weather have a lot in common with the Long Range Traverse—I couldn't help questioning whether it was all worth it, much less whether I should be encouraging *Times* readers to follow suit. "Why," I wondered, "on our preciously rationed vacation days, were we here?"

And yet there we were once again, in the summer of 2022, stumbling semiblindly through the mist and muck of the Long Range Mountains—this time with our kids, who hadn't really signed up for any of this. After a few more wrong turns, Lauren and I accepted the inevitable and started looking for a patch of rock flat and puddle-free enough to pitch our tent. I lay awake that night calculating and recalculating how long it would take us to complete the hike and how much spare food we had, and contemplating yet again what exactly had drawn me here.

This wasn't my first sleepless night of the summer. Even back home, nestled in the comfort of my pillowtop mattress, I'd been regularly finding myself awake and staring at the ceiling in the small hours of the morning. My mind would spin through the various dramas of the week, and then zoom out to broader existential musings about my life path. Two years into the pandemic, I certainly wasn't alone in contemplating my choices. But I'd been stuck in this loop since before the pandemic even started.

In 2018, I published a book called *Endure*, which explored the evolving science of human endurance. It was the culmination of a decade of reporting, during which my journalistic endeavors had become ever more narrowly focused on that specific topic. I'd started my freelance career reporting for a wide variety of publications on physics, jazz, accounting, travel, philosophy, and anything else that piqued my curiosity. But by 2018, I was a regular columnist for *Outside*, writing about the science of endurance, having moved there from a columnist gig at *Runner's World*, where I had begun writing about the science of endurance in 2012, while continuing to moonlight as a columnist at *Canadian Running* and the *Globe and Mail*, where in both cases I wrote mostly about the science of endurance.

Endure did unexpectedly well. It scraped briefly onto the *New York Times* bestseller list, and it positioned me perfectly to brand myself as "the science of endurance guy" and milk that role for the rest of my working life. Speaking invitations flowed in; doors opened at magazines that I'd always dreamed of writing for. To the extent that my younger self had ever managed to conjure up a vision of a dream career (not including winning the Olympics as a runner), this was it. But something didn't feel quite right. My decade of reporting for *Endure* had been a period of continual discovery, as I learned about new and new-to-me developments in biology, physiology, psychology, and other disciplines. By 2018, though, I was mostly caught up with the current state of knowledge. A future of reporting exclusively on those same topics would mean waiting for rare incremental advances and rehashing ideas I'd already written about. The spark of learning something new was gone.

The obvious move after a book like *Endure* is to get busy on a follow-up. But instead I found myself tugged in other directions, none of which sustained my attention for long. Weighing on my mind was my prior history of career swerves. I had started out studying physics. After submitting my PhD thesis, at age twenty-four, I decided to go all-in as a middle-distance runner and train full-time in an attempt to qualify for the Olympics. A little over a year later, I checked my world ranking and my bank balance and decided to be a physicist after all. I took a postdoctoral research position in the National Security Agency's quantum computing group, working out of a lab affiliated with the University of Maryland. It was fun and intellectually rewarding, but two and a half years later, at age twenty-eight, I left my postdoc to start a master's degree in journalism at Columbia University. Was I astutely following my interests? Or, I sometimes wondered, was I a dilettante, chasing whatever shiny new object caught my eye instead of sticking to my original challenges?

What I worried about a decade and a half later, as I wandered in the post-*Endure* wilderness, was that I might be repeating this pattern: setting an audacious goal, spending years working tirelessly toward it, and then, once success was within reach, walking away to pursue something completely different. A decision like that might seem quixotic when you're twenty-eight, but in your midforties it starts to look pathological. So there was an insistent voice in my head exhorting me to exploit all the effort I'd put in to become "the science of endurance guy." And there was another, quieter voice reminding me that the seemingly irrational decision to explore a journalism career had led to the most rewarding professional years of my life, and that following that urge one more time might pay off again. In other words, I eventually realized, I was on the horns of a ubiquitous and exhaustively studied metachoice that researchers call the explore-exploit dilemma.

In 1991, a professor named James March, at Stanford University's Graduate School of Business, published a paper called "Exploration and Exploitation in Organizational Learning." March was a prolific and influential scholar, as well as a polymath: he published poetry and produced films about the leadership lessons of *Don Quixote* and *War and Peace*. Starting in the 1950s, his work with economics Nobel Prize–winner Herbert Simon and others brought nuance and complexity to the study of corporate decision-making. More than a few scholars believe March should have shared Simon's Nobel.

March's 1991 paper highlighted the fundamental tension between what he dubbed *exploration*, encompassing "search, variation, risk taking, experimentation, play, flexibility, discovery, innovation," and *exploitation*, encompassing "refinement, choice, production, efficiency, selection, implementation, execution." You can exploit the knowledge and resources you already have, or you can explore in search of an outcome that is uncertain but might turn out to be better. You can devote your corporate resources to churning out widgets as cheaply and efficiently as possible, or you can devote them to inventing a sprocket that will make widgets obsolete. But in a world of finite resources, you can't give your all to both at the same time. You have to choose. March's main argument was that the delayed and uncertain rewards of exploration mean that organizations tend to systematically underinvest in it.

The paper also had an unintended side effect: the "explore-exploit" terminology caught on, crystallizing a concept that researchers in various academic silos had been grappling with in their own specialized languages. In the years that followed, mathematicians who had been toiling for decades on complex optimization algorithms realized that they were addressing the same fundamental questions as economists and business thinkers like March, as well as evolutionary biologists studying human migration paths, ecologists examining animal foraging patterns, neuroscientists decoding the brain's decision circuitry, computer scientists teaching machines to learn, and psychologists and philosophers trying to understand why we want what we want.

As an ex-physicist, I was a sucker for the mathematical approach to explore-exploit decisions. I wasn't naive enough to imagine that I could plug a few details about myself into an equation and get quantifiable advice about my next vacation or my next book. But I began to see explore-exploit dilemmas all around me: in the tug-of-war between my long-standing love of running and my emerging interest in rock climbing; in the music I chose to stream; in the friendships I chose to maintain, neglect, or initiate; in the investment choices I made with my retirement savings; in the search for alternatives in my writing to familiar clichés and overused adjectives. And as I dug into a century's worth of progress on exploring algorithms, I found insights that helped me think through my dilemmas. For instance, the math is pretty clear about the lesser value of pure exploration in your forties compared to your twenties. "Your horizon is getting shorter," a cognitive scientist at Georgia Tech, Robert Wilson, explained to me. There's less time to reap the delayed benefits of a new path, and there's plenty of evidence that humans get worse at exploring and do it less often as they age—though that doesn't mean I should lean into this decline.

The quantitative approach also offers some clues about a thornier question: the *why* of exploring. If you want to teach a computer to learn about the world, it's helpful to program it with an "uncertainty bonus." A navigation algorithm, for example, might suggest route choices based on what has minimized travel times for similar trips in the past. But the algorithm will generate better results if you incentivize it to also check some possibilities that it hasn't sampled recently, just in case there are new road upgrades or favorable traffic conditions there. It turns out that, in many real-world contexts, we seem to include exactly this sort of uncertainty bonus in our decision-making calculus. As we'll see, we pick new restaurants based mostly on how good they're reputed to be—but if all else is equal, and sometimes even if it's not, we opt for ones we know less about.

There's a reason we're wired this way: exploration works. In recent years, bestselling books have extolled the power of good habits, which are exploit decisions in a hyperpure form. Good habits are certainly important: by one estimate, about 45 percent of our actions in a given day are habitually driven. But it's easy to get stuck in suboptimal routines.

Even regular commuters, who retrace the same route twice daily, often turn out to be taking slower or less pleasant routes than alternatives they haven't tried. Exploration, in this sense, is the antihabit, and it has paradoxical effects: a single instance of exploring will likely yield a worse-than-usual outcome, but the collective effect of repeatedly breaking free of your usual routines will be better outcomes—a faster commute, for example—in the long term. By breaking old habits, the uncertainty bonus helps you build better ones.

The lure of hiking the Long Range Traverse also starts to make more sense when you think in terms of an uncertainty bonus. For starters, neither Lauren nor I had ever been to Newfoundland. We've hiked extensively in the Rockies, so we know exactly how beautiful they are. Uncertainty bonuses are encoded in our brains, in part, with "reward prediction errors": you get a shot of dopamine not because something is *good*, but because something is *better than expected*. That's why, for a certain type of person, a decent view in Newfoundland might trump a jaw-dropping vista in Banff; or fresh pakoras in a Delhi back alley make it worth skipping the Taj Mahal. It's also, in part, why people who are addicted to drugs need a progressively bigger dose to get the same high.

The most potent source of uncertainty in the Long Range Mountains, though, isn't the view; it's the hike itself, with no prescribed route and no trail markings. These days pretty much every travel experience—including, for better or worse, the Long Range Traverse—is documented on someone's travel blog. For planning purposes, these blogs are amazing resources: you get a better sense of how long a route will take, what conditions you're likely to encounter, what gear you'll need, and so on. The danger, though, is that the trip then goes exactly as you predicted. You won't discover halfway through a hike that you really should have brought crampons, which is great. But you also won't get blown away when you turn a corner and discover a hidden waterfall: you've already seen the pics.

Choosing your own route through the mountains reinjects some of that uncertainty—the possibility of prediction error—into the experience. By necessity, our first camping spot along the Long Range

Traverse was nowhere near any of the spots we'd read about or scouted. There was no source of water nearby, so I had to bushwhack back down the slope for ten minutes until I found a little rivulet that was clear and deep enough to fill our bottles. We had to scour far and wide to find rocks to hold our tent down in the wind, because no prior campers had left a convenient pile. These added challenges were inconvenient, but they also reinforced our sense that we were discovering this world afresh rather than simply following a well-trodden conveyor belt past some prepackaged scenic viewpoints—an illusion, perhaps, but an engaging one.

Is a hike through a national park really "exploring"? One view is that true exploring involves venturing into territory where no human has preceded you: if there are footprints, you're not exploring. Alternatively, you could argue that exploring is simply another word for trying something new: if the TV show you're watching gets boring and you change the channel, you're exploring what else is on the airwaves. Neither of these definitions really captures what the concept means to me. The Latin word *explorare* meant to reconnoiter, inspect, or investigate. It was formed from *ex* (from or out of) and *plore* (to wail or lament); the original meaning is thought to have been "to scout the hunting area for game by means of shouting." That's not quite what I mean either, but there's the kernel of something important there: you're seeking information rather than just novelty.

Meaningful exploration, I will argue, involves making an active choice to pursue a course that requires effort and carries the risk of failure—what the mythologist Joseph Campbell called "a bold beginning of uncertain outcome." Most importantly, it requires the embrace of uncertainty, not as a necessary evil to be tolerated but as the primary attraction. If you're given a choice between being shot or being banished into the jungle, you choose the jungle to maximize your odds of survival. Exploring, by contrast, is heading into the jungle when your alternative is being an accountant. The stakes may be great or small, and the undiscovered country may be literal or metaphorical, but by choosing the uncertain option you're seizing an opportunity to learn about the world. It might even be the murky boundaries of your own capacities and limits that you're seeking to discover—a goal that maps nicely onto endeavors like running a marathon ("the great suburban Everest," as London Marathon founder Chris Brasher put it) or hiking in a national park.

We did, in fact, make it to the end of the Long Range Traverse, more or less on schedule and with a few scraps of food left in our packs. Compromises were made. We skipped a side trip up to the peak of Gros Morne Mountain. And starting on the second day, I began to rely increasingly on the GPS waypoints I'd loaded onto my phone from the Parks Canada site. My original intention was to have them available as a safeguard if we became unsure of our position. Instead, I ended up hiking most of the way with my phone in my hand, using the digital topo map and waypoints to guide us in real time. Something was lost in the process, and I knew it. But our margin of safety—and the kids' tolerance—had worn too thin to risk any long detours or backtracks.

After the hike was done, and the celebratory mooseburgers and moose-size ice cream cones had been consumed, we spent the next few days driving up the west coast of Newfoundland, to its very northern tip. There, on a characteristically drizzly gray day, we poked around a boggy green meadow dotted with grassy mounds. Directly to the north of us, beyond a stern and rockbound coast, was the open water of Iceberg Alley and the Labrador Sea. Next landmass: Greenland.

In the spring of 1960, a Norwegian explorer and adventurer named Helge Ingstad began a painstaking search in the seaside town of Newport, Rhode Island. Ingstad had spent years combing through the ancient Icelandic Sagas for clues about the location of Vinland, the short-lived Viking settlement supposedly founded by Leif Erikson, traveling from Greenland around 1,000 AD. Ingstad was looking for geographical clues that would match the description in the Sagas: a grassy meadow, a small river leading to an inland lake, a mountain whose ridgeline looked like the overturned keel of a ship. And better yet, he was looking for ruins that would confirm the presence of pre-Columbian Norse settlers. Newport had a "Norse" stone tower, but it turned out to be an eighteenth-century chimney.

Over the next few months, Ingstad traced the coastline north through Cape Cod, Boston, New Hampshire, Maine, Nova Scotia, and eventually Newfoundland. Everywhere he asked the same questions about landmarks and ruins, but it wasn't until he reached the northern tip of Newfoundland that he finally got the answer he was looking for. "Yes, I have heard of something like that," a man in the tiny fishing village of Raleigh told him. "Over at L'Anse aux Meadows. But you need to talk to George Decker."

L'Anse aux Meadows, at the time, was an even smaller fishing village with just thirteen families, accessible only by boat. It had a small river, an inland lake, a keel-shaped mountain, and open meadows where George Decker grazed a few sheep and cows, and where the local children played among what they called the "Indian mounds." Ingstad's ride—a medical mission's small boat carrying a nurse along the northern coast to vaccinate children in remote outports—was soon leaving, but he made plans to return the next year with his wife, a trained archaeologist named Anne Stine Ingstad. In succeeding years, the Ingstads led the excavation of what is now generally assumed to be Leifsbudir, or "Leif's camp," the heart of the Vinland settlement. To help dig, they hired a few of the locals, including a young boy named Clayton Colbourne.

When we visited L'Anse aux Meadows in 2022, Clayton Colbourne—now a trim septuagenarian with a thick white beard—was our tour guide. He was hired when Parks Canada took over the site in 1973, helping to build the replica sod houses where visitors to the national park now hobnob with costumed Vikings. In his current role, he finds that visitors are almost as fascinated by his tales of growing up in L'Anse aux Meadows in the 1950s as they are by the saga of the Viking settlement and its rediscovery. As we wandered through the remains of the settlement—a large leader's hall, huts for the crew, a shed for boat repair, a smelting hut that appears to have been used just once—his patter jumped back and forth between the distant past and a more recent past that, to us, was almost as foreign. "I used to play on these mounds as a kid," he said. "We just figured they were Indian remains."

Along the boardwalk leading across the bog from the archaeological site back to the visitor center, Colbourne paused at a giant two-piece sculpture looming like an arch above the path. The curves and whorls of the three-thousand-pound bronze monument evoke billowing sails, crashing waves, and, more abstractly, two hands reaching toward each other. *The Meeting of Two Worlds*, by Newfoundland sculptor Luben Boykov and Swedish sculptor Richard Brixel, was unveiled in 2002 to symbolize the closing of a giant loop. After humans migrated out of Africa, some went east through Asia and across the Bering Strait into the Americas; others went west through Europe. "This is where they came full circle," Colbourne said. "This was their first meeting in a hundred thousand years."

We can quibble about the exact dates. Patterns of early human migration are complex and still the topic of vigorous academic debate. But the idea—the symbolism of this massive monument, perched on the edge of an angry sea in one of the remotest corners of the continent—stopped me in my tracks. Scattered around the Norse site are excavated fire pits and tent rings, along with debris from toolmaking left behind by at least five different Indigenous groups dating back as far as five thousand years ago. The Norse weren't the only ones who made intrepid journeys to reach this spot. In fact, the people who were there waiting for them had made even more improbable voyages, through harsher environments, with much simpler technologies—but spurred, perhaps, by the same unnamed urge.

Standing beneath Boykov and Brixel's archway, the dilemmas that had been keeping me up at night—my masochistic fixation on the vacation itinerary less traveled; the recurring allure of a freshly trodden career path—began to feel like part of a much larger human story. Like my far-flung and long-forgotten ancestors, and like everyone else on the planet, I was born to explore. That exploration can take many forms for different people, and it has changed—and will continue to change—across my lifespan. The great age of geographical exploration has mostly passed, at least here on Earth, but exploration in a broader sense has never been more important as we confront destabilizing shifts in technology, society, and climate.

I've come to believe that the drive to explore can be both a source of meaning in our lives and a spur for growth. What makes exploring hard—the uncertainty, the struggle, the possibility of failure—is, at least in part, what makes it rewarding. That doesn't mean pushing onward to see what's around the next corner or over the next ridge is always the right choice, though. Letting your exploring circuitry take the reins can also leave you starving in the jungle, stranded on an ice-jammed ship, or staring endlessly at the flickering screen of your phone. To harness the power of exploring, then, we need to understand why we're drawn to the unknown, what we're seeking there, and how we can do it better.

Part I

WHY WE EXPLORE

1

THE GREAT HUMAN EXPANSION

On the thirtieth day of the voyage, Mau the navigator predicted they would find land the next day. Since setting out in their slender, double-hulled canoe, they'd been alone on the ocean, traveling more than 2,500 miles without landmarks or navigational aids other than those provided by the waves, wind, and sky. There were seventeen of them crammed into the sixty-foot-long vessel, along with a pig, a dog, and two chickens. Alongside their provisions, which they supplemented by fishing and collecting rainwater, they had plant cuttings, seedlings, and roots, wrapped in moss and a type of cloth made from tree bark to protect them from salt water—all the things they would need to establish a new settlement.

The voyage had been hard. They'd been battered by gales, becalmed in the oppressive heat of the doldrums, and soaked by rain squalls as they tried to sleep on the open deck. The waterproof compartments in the hulls had taken on water; a rope anchoring one of the boat's two masts had snapped, putting dangerous strain on the mast itself. Hundreds of pounds of taro root they'd brought as a staple food turned out to be full of maggots. But shortly after Mau's pronouncement, someone spotted small white birds skimming over the waves in the distance. They were fairy terns, which seldom range farther than about thirty miles from the coral islands where they nest. Later in the afternoon, the rhythmic rise and fall of the boat began to taper off, suggesting that something beyond the horizon must be blocking the steady swell produced by the trade winds.

Shortly before dawn the next morning, as Mau had predicted, they came to an island.

Before we can wrestle with the big and gnarly questions—Why do we explore? What do we get out of it? How do we know when it's a good idea and when it isn't?—we have to consider a more basic one. *Do* we explore? Is there some fire that burns within us, goading us to find out what lurks around the next corner, or what lies across the uncharted ocean?

You might think the answer is obvious. After all, we *have* crossed uncharted oceans. But that doesn't prove that humans find uncharted oceans irresistible. Humans have spread across the globe, but so too have other animals and plants. Young wolves, for example, sometimes travel immense distances into lands they've never seen before. "In evolutionary terms, they are 'searching for their own territory,'" says Rebecca Wragg Sykes, a British archaeologist who studies Neanderthals. "But do they actually feel that level of motivation? Or are they simply attracted to exploring the next piece of ground, or crossing to that intriguing-looking riverbank, or trying to reach that far tree on the ridge?"

Similarly, tree seeds spread by blowing in the wind, but no one romanticizes the noble exploratory urge of the acorn. You can model how various species of tree have expanded their range by using the mathematical laws of diffusion, first formulated in the 1850s by the German physicist and physiologist Adolf Fick. And scientists like Joaquim Fort, of the University of Girona in Spain, use the same equations to model how human populations, and even particular technologies and ideas, diffuse into new territory. "Honestly," Fort told me, "I see no reason to believe that humans spread differently than other animals or even vegetables."

So why *did* Christopher Columbus cross the Atlantic? Well, according to his published journals, it's because he wanted to convert the princes of India to the holy Christian faith. Oh, and also because the king and queen of Spain promised that he could be "perpetual Viceroy and Governor of all the islands and continents that I should discover and gain ... and that my eldest son should succeed, and so on from generation to generation for ever." In other words, the hypothetical urge to explore was mixed with other motives. One of the first scientists to attempt a formal theory of exploratory behavior, the psychologist Daniel Berlyne, attributed it in 1960 to a mix of "internal predisposing factors," biological utility, and the pursuit of rewards—along with what he termed "ludic behavior," or play. The same mix of motivations is present in later explorations. Polar explorers were seeking fame and glory. Neil Armstrong and his crew were trying to beat the Soviets. George Mallory, whose famous quip to a *New York Times* reporter about his motivations for trying to climb Mount Everest—"Because it's there!"—remains the pithiest formulation of the adventure-for-its-own-sake spirit, was motivated by patriotic pride.

Even today, the question of whether we're wired to explore remains both relevant and contested. Should we invest vast resources and take deadly risks to pursue the exploration—and perhaps the colonization—of outer space? "Exploration is in our nature," Carl Sagan claimed. "We began as wanderers, and we are wanderers still. We have lingered long enough on the shores of the cosmic ocean. We are ready at last to set sail for the stars." But the argument that we should head to Mars because that's just what humans do isn't very convincing, according to J. S. Johnson-Schwartz, a philosopher who studies the ethics of space exploration. The problem, in her view, is that space advocates have simply assumed that we have an innate drive to explore, rather than marshaling the evidence to demonstrate it. So let's try to fix that.

In the next three chapters, I'll present three lines of evidence that suggest we're natural-born explorers. The first is anthropological, assessing why and how humans spread across the globe. The second is biological, tracing how exploration has affected our genes and vice versa. And the third is neuroscientific, drawing on a newly ascendant theory of the brain called predictive processing. Taken together, they make a compelling case that we really do find it intrinsically rewarding to seek out the unknown.

There are two great exploring narratives in our past, according to the historian Felipe Fernández-Armesto. The first is how human cultures emerged from the African savannah and, starting in earnest about fifty thousand years ago, spread to every habitable corner of the planet. The second is how, in more recent times, those far-flung cultures rediscovered each other. We know a lot more about the second story—the Leif Eriksons and Marco Polos and Christopher Columbuses—because we have diaries and histories and physical artifacts from those voyages. But some of the most intriguing hints of a drive to explore show up in the fragmentary clues we've pieced together about the first story, the period some historians refer to as the Great Human Expansion.

The current thinking is that anatomically and behaviorally modern humans—the ones whose descendants settled the rest of the world—left Africa and the Near East around fifty thousand years ago. They weren't the first ones to make the trip: successive waves of archaic human species like *Homo erectus* spread into Asia and Europe starting almost two million years ago. Neanderthals made it there too, as did other now-extinct human populations like the Denisovans and the so-called hobbits discovered on the Indonesian island of Flores. In fact, our own species, *Homo sapiens*, also migrated out of Africa several times, perhaps as early as 270,000 years ago. These early emigrants were *anatomically* modern—that is, you could shave one, dress him in a suit, and successfully pass him off as one of us, at least from a distance. But they weren't yet *behaviorally* modern: they didn't paint the walls of their caves, build elaborate hearths for their fires, ceremonially bury their dead, or set off on expeditions to the North Pole.

This concept of behavioral modernity is a fuzzy one, and it's clear that some of its hallmarks, like symbolic thinking, were gradually emerging well before the final decisive migration of *Homo sapiens* from Africa. Still, there was an inflection point. The archaeological evidence for behavioral modernity shifts from a trickle to a flood around fifty thousand years ago, right around the time that humans with these traits began to sweep across the globe. Svante Pääbo, who pioneered the analysis of ancient DNA from Neanderthal remains (for which he earned a Nobel Prize in 2022), calls this modern population "the replacement crowd," because their spread across Europe and Asia coincided with the demise of the human populations that preceded them, give or take a little interbreeding. And unlike previous waves, they didn't stop at the outer boundaries of Eurasia.

Over the next fifty thousand years, the replacement crowd ventured into unpopulated lands as far as twenty thousand miles away from their starting point, leaping across the so-called Wallace Line that separates Asia from Oceania, crossing from Siberia into Alaska, and trekking all the way down to the southern tip of South America. Fernández-Armesto's first narrative draws to a close about eight hundred years ago with the settlement of New Zealand, by which point pretty much every habitable place on the globe was occupied, from the remote Australian outback to the frigid steppes of Greenland.

There's something unique about this kudzu-like thirst for new territory, Pääbo argued in a *National Geographic* article—something distinctively human. "No other mammal moves around like we do," he said. "We jump borders. We push into new territory even when we have resources where we are. Other animals don't do this. Other humans either. Neanderthals were around hundreds of thousands of years, but they never spread around the world. In just fifty thousand years we covered everything. There's a kind of madness to it. Sailing out into the ocean, you have no idea what's on the other side. And now we go to Mars. We never stop. Why?"

In the late 1950s, a young surfer from San Diego named Ben Finney was working on a master's degree in anthropology at the University of Hawaii. His thesis was about surfing in Polynesian culture, but his adviser, the folklorist Katherine Luomala, gave him a copy of a seemingly unrelated book. "I don't like this book," she told him, "and I want you to read it."

The book was called *Ancient Voyagers in the Pacific*, by Andrew Sharp, an acerbic civil servant and part-time historian from New Zealand. In it, he tackled one of the great mysteries in the history of human migration: How on earth did the Polynesians get to Polynesia? Its thousand-odd islands are scattered across a vast triangle of ocean between Hawaii, New Zealand, and Easter Island, with a total area comparable to all of North America. You don't wind up on Easter Island because you take a wrong turn, or because things are getting a little crowded in your neck of the woods. And yet every habitable island, no matter how seemingly desolate or isolated, was already inhabited by the time European seafaring prowess had advanced enough to explore the region. The settlement of Polynesia was the very last chapter of Fernández-Armesto's first narrative, taking place after the second story had already begun elsewhere in the world. And its unique geography—its epic water crossings, in particular—make it an ideal laboratory to test theories about the instinct to explore.

The Spanish navigator Álvaro de Mendaña was the first European to land in Polynesia, arriving from Peru to the Marquesas Islands in 1595. (Ferdinand Magellan had sailed through the region on his three-month crossing of the Pacific in 1521, but he didn't spot any islands except a couple of small uninhabited atolls.) Mendaña's navigator, Pedro Fernandez de Quiros, puzzled over where the islands' inhabitants could have come from, given his impression that they were a "people without skill or the possibility of sailing to distant parts." He figured that there must be a chain of islands just over the horizon, leading down to a great southern continent—but thanks to a mutinous crew and a badly mistaken belief that they were almost finished crossing the Pacific, Mendaña sailed on without investigating further.

Subsequent explorers were just as baffled. The Dutch navigator Jacob Roggeveen, who "discovered" Easter Island in 1722, speculated that its inhabitants were "descendants of Adam" who had always been there, since it was impossible to understand how they might have arrived from elsewhere. Half a century later, Frenchman Julien Crozet astutely noticed the similarities in language and appearance between Polynesian islanders as far apart as Tahiti and New Zealand. His explanation: the whole region must once have been an above-ground continent, "of which volcanic shocks have left us only the mountains and their savage inhabitants."

It eventually occurred to curious Europeans that they could simply ask the Polynesians where they had come from. The early explorers, even when they'd thought to ask, hadn't learned the local languages or understood the cultures well enough to understand the seemingly inscrutable answers they received. But by the 1820s, as more Europeans settled in the region, language barriers were less of a problem. On island after island, they heard stories of epic voyages of discovery in the not-so-distant past, mixing clearly supernatural elements with specific details about the who, when, how, and why of various migrations. Among the first to knit these disparate threads into a cohesive narrative was a Swede named Abraham Fornander, who ran away to sea as a nineteen-year-old in 1831 (after, by some accounts, falling unhappily in love with his mother's youngest sister) and eventually, after a decade as a harpooner on whalers in the Pacific, washed up in Hawaii. There, he married a Hawaiian woman and became a citizen by swearing allegiance to King Kamehameha III, and spent the subsequent decades collecting traditional oral histories and detailed genealogies from around the islands.

In his 1877 masterwork, *An Account of the Polynesian Race, Its Origin and Migrations, and the Ancient History of the Hawaiian People to the Times of Kamehameha I,* Fornander laid out a multistage journey. The ancestors of Polynesians had, he claimed, started in the foothills of the Hindu Kush (a misguided theory based on imagined linguistic similarities); spread to Indonesia and the Philippines at some indeterminate point in the past; island hopped to Fiji shortly after the birth of Christ; continued on to Hawaii by around 500 AD; then settled the rest of Polynesia, including New Zealand, between about 1,000 and 1,400 AD. Subsequent scholars tweaked the details of Fornander's account—it is, after all, an inexact science to estimate dates by counting the dozens of generations in an oral family tree that ultimately leads back to Wakea, the Sky Father, and Papa, the Earth Mother, residing on the Indonesian island of Gilolo two thousand years ago. But in its rough outlines (aside from the bit about the Hindu Kush), this was the predominant view of Polynesian origins when Ben Finney was in grad school—and notably, it depended on a belief that early Polynesians had the skills, technology, and desire to deliberately travel hundreds or even thousands of miles across the open ocean. They were, as one of Fornander's scholarly heirs put it, "Vikings of the Pacific."

This version of history was reinforced by a pan-Polynesian culture that seemingly prized intrepid voyagers. Joseph Banks, the naturalist who accompanied Captain James Cook's first explorations of Polynesia from 1768 to 1771, noted somewhat disbelievingly that the Polynesians claimed to take their flimsy-looking double-hulled canoes on "very long voyages, often remaining out from home several months, visiting in that time many different Islands of which they repeated to us the names of near a hundred." In 1837 (two years before he was eaten by cannibals, as it turned out), the English missionary John Williams recorded that in the Society Islands it was "an object of ambition with every adventurous chief to discover other lands." Around the same time, an American ethnologist noted that Polynesians seemed to be "cosmopolites by natural feeling," surprisingly eager to hop on board passing ships "for no purpose but to gratify their roving disposition, and their desire of seeing foreign countries."

Even as the influence of European and other outside cultures began to encroach on the region's traditions, Polynesian wanderlust remained a familiar trope. Polynesian deckhands became a familiar sight on ships around the world—think of Queequeg, the harpooner in *Moby Dick*. To Raymond Firth, an ethnologist from New Zealand who visited the isolated Melanesian island of Tikopia in the 1930s, its inhabitants were still "fired by the lust for adventure and the desire to see new lands" as they set out in their canoes. Even *Moana*, the 2016 Disney movie, plays on the theme. "See the line where the sky meets the sea?" Moana sings. "It calls me."

But it was precisely this image that Andrew Sharp set out to puncture in *Ancient Voyagers in the Pacific*, the book that Ben Finney's thesis adviser gave him. Sharp simply didn't believe that early Polynesians were capable of deliberate long voyages across the open ocean. Their boats, which lacked keels to sail against the wind and, on sparsely forested islands, were sewn together from scraps of wood, weren't good enough. More important, they had no navigational technology: no sextants, no compasses, no means of keeping accurate time, none of the advances that eventually allowed Europeans to accurately track their position when land wasn't in sight. Polynesian navigators used the sun and stars—but, Sharp pointed out, those are often obscured. They also relied on prevailing winds and currents, but these too are changeable. Beyond a radius of about three hundred miles, Sharp declared, deliberate Polynesian voyages of exploration or colonization, much less return travel between islands, were impossible.

Sharp's book was immediately and enormously controversial—"One of the most provocative studies ever in Pacific history," the New Zealand historian Kerry Howe later judged it. "I felt that Andrew Sharp had set out with the purpose of discrediting the achievements of our Polynesian ancestors," the Maori scholar Pei Te Hurinui Jones wrote in 1957. Many historians agreed that Fornander and his heirs had been too credulous in trying to extract reliable dates and migration routes dating back thousands of years from confusing and contradictory oral histories. But on the question of whether those deliberate voyages had *ever* happened, he encountered fiercer pushback.

The most likely explanation, in Sharp's view, was accidental voyaging: canoes pushed off course by storms, or exiles forced into one-way voyages with unknown destinations. He wasn't the first to suggest this. In fact, he cited the authority of none other than Captain Cook. On his first voyage, Cook met a learned priest and navigator named Tupaia in Tahiti. Tupaia explained to Cook and his crew how the Tahitians navigated by the sun, moon, and stars, and drew a remarkable map showing the position of seventy-four islands within a four-thousand-mile radius, most of which he or his relatives had visited. When, after three months, Cook sailed away from Tahiti, Tupaia joined him on board, guiding the Europeans and interpreting for them on other islands. The fact that Tupaia could be understood as far away as New Zealand convinced Cook that Polynesians were all part of the "same Nation"— and, by extension, that their voyaging skills were sufficiently advanced to explain how they spread "from Island to Island" across the vast expanses of the South Pacific.

But a decade later, on his third voyage to Polynesia, Cook had second thoughts. On the remote islet of Atiu, he met three Tahitians, the survivors of a twenty-person sailing trip that had been blown off-course on a short inter-island jaunt twelve years earlier and drifted more than six hundred miles until being rescued by the Atiuans. "The application of the above narrative is obvious," Cook wrote in his journal. "It will serve to explain, better than a thousand conjectures of speculative reasoners, how the detached parts of the earth, and, in particular, how the islands of the South Sea, may have been first peopled; especially those that lie remote from any inhabited continent, or from each other."

In Sharp's view, Cook was right the second time. Sharp's thesis—and his obnoxious insistence that he could "prove to any perceptive mind" that Polynesian long-distance navigation was impossible—provoked plenty of debate. It also spurred some direct responses, including an ambitious computer simulation of Pacific drift patterns. Starting in 1964, a team of researchers fed thousands of tables of wind and current data into an early Ferranti Atlas computer that filled two floors of a building at the University of London, then simulated more than a hundred thousand possible drift voyages from different starting points around Polynesia. Some routes, like Tonga to Fiji, were indeed likely; but others, in particular those leading to Hawaii, New Zealand, and Easter Island—the corners of the Polynesian Triangle—were essentially impossible.

Accidental drift voyages, in other words, couldn't explain the settlement of Polynesia. But the researchers also tested another possibility. What if voyagers set out with what the scientists termed "intent to explore," sailing into the open ocean *in a specific direction* rather than simply being blown wherever the wind and currents took them? Given the limitations of double-hulled canoes, the simulations assumed that they couldn't sail directly into the wind, but could at least hold a course perpendicular to it. With this additional twist, the computer returned a more favorable verdict: Polynesians could indeed have sailed to all possible islands, but only by deliberately setting out to explore the unknown.

Sharp was unmoved by these arguments, or indeed by any contrary opinions. But Ben Finney, the young graduate student in Honolulu,

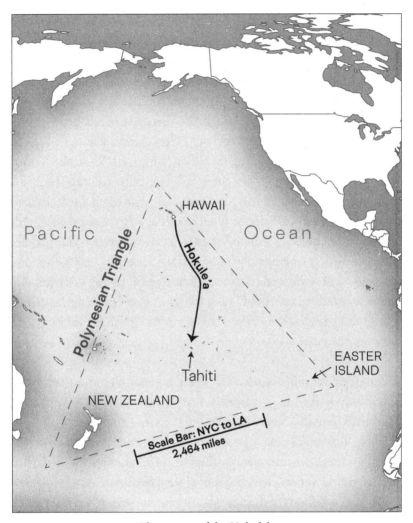

The voyage of the Hokule'a

had been contemplating a more direct response to Sharp's claims ever since reading his book in the late 1950s. The brainwave came to him one afternoon after class, Finney later recalled, "while flying before the trades in a racing catamaran, a modern descendant of the ancient voyaging canoe." If Sharp said sailing between Hawaii and Tahiti with Polynesian technology and knowledge was impossible, why not prove him wrong by doing it?

In 1965, while teaching at the University of California Santa Barbara,

Finney and his wife, Ruth, began building a forty-foot double canoe based on eighteenth-century drawings of Hawaiian sailing vessels. Despite its lack of keel or centerboard, they eventually found that it could sail to within about 75 degrees of a fifteen-knot wind—better than the ninety degrees assumed by the computer simulations. To refute Sharp with a long-distance voyage, though, they needed a bigger boat. In 1971, Finney took a job at the University of Hawaii, and a few years later cofounded the Polynesian Voyaging Society with Herb Kane and Tommy Holmes. Raising money, designing a historically accurate boat, and unearthing lost navigational knowledge all proved to be formidable challenges—but in 1975, to the ritual chants of *E Hoʻomakaukau!* ("Make ready!) and *E alulike!* ("Pull together!"), a crew of twenty-four paddlers and a steersman ceremonially launched the *Hokuleʻa*, a sixty-foot double canoe with twin crab-claw sails modeled on petroglyphs found in a remote lava field on Hawaii's Big Island. A year later, after a series of shorter test voyages, the *Hokuleʻa* set out from Hawaii for Tahiti.

The mathematical approach to studying population movements works remarkably well in most contexts. People and plants alike tend to expand their ranges according to much the same rules that govern, for example, how epidemics spread. But there is one situation where the equations break down: water crossings. In 1876, a botanist with the British Geological Survey named Clement Reid began collecting fossilized seeds from a famous archaeological site in Norfolk called the Cromer Forest Bed. Reid was one of the first scientists to systemically study the distribution of fossil seeds, and two decades later he laid out what became known as Reid's Paradox. Oak trees, in order to cross the English Channel from Europe and repopulate Britain after the ice age that ended 11,500 years ago, would have had to travel at least six hundred miles. To spread that far simply by shedding acorns onto the nearby ground, he estimated, would take a million years.

To get a more realistic estimate, you have to assume that some fraction of seeds travel surprisingly long distances, supercharging the pace of range expansion. There are plenty of ways that can happen for small seeds: blowing in the wind, stuck on the feet or in the feathers of migrating birds, tangled in the fur of passing mammals. Larger seeds like acorns are trickier, especially when they have to leap across a marine barrier. Reid, in his 1899 treatise *The Origin of the British Flora*, imagines some fairly intricate scenarios for the arrival of oak trees: "an ox, a deer, or a horse falling over the cliffs of France would tend to drift with the prevalent south-west wind till it was thrown upon the English Coast, where wolves and foxes would pull it to pieces, dragging the remains beyond the reach of the sea, and perhaps burying parts, with the undigested vegetable food still contained in the stomach."

That's clearly not how humans crossed the English Channel, but Reid's Paradox shows how tricky it is to draw conclusions from the fossil record about why and how a given species—even a simple oak tree—might migrate. Joaquim Fort and others have found that the first human farmers spread across Europe and Asia at a rate of about one kilometer per year. But the speed of population diffusion jumps tenfold, to ten kilometers a year, when post–ice age humans were migrating along water routes. There's a lively debate among scientists about what, if anything, we can infer from such patterns. Were early water crossings to places like Crete and the Indonesian island of Flores—settled, in the latter case, at least a million years ago across a gap that likely measured tens of miles—accidental or deliberate? Was the necessary planning complex enough to require language or other forms of symbolic communication? Were compound tools needed to build the boats?

Beyond a certain scale, though, there's no more ambiguity. To reach Australia from southeast Asia some fifty-five thousand years ago, humans had to island-hop from the Asian continental shelf to the Australian one across gaps of at least fifty-five miles, by one estimate. That's too far to see your destination across the water. In fact, one view is that the water crossings that led to Australia provide the earliest indisputable archaeological evidence of behavioral modernity in the world. You can argue about the true meaning of cryptic cave paintings and primitive fireplaces, but there's simply no way humans could have made it to Australia without language, planning, and technology. And a fifty-five-mile voyage is nothing compared to the trips of more than three hundred miles that were required to settle Polynesia.

There's something deeper here too. These sea voyages, according to Florida State University anthropologist Thomas Leppard, demand *intention*. "It is a complex enough task to design, build and utilize a raft to cross oceanic distance," he writes; "quite another to stand on a beach, staring at the horizon, imagining distant worlds to which the raft might be paddled and on which life might be lived more fully and satisfactorily."

Finney's voyage from Hawaii to Tahiti was predicated on the idea of a pan-Polynesian culture that glorified "staring at the horizon, imagining distant worlds," and that had the know-how to sail back and forth between distant islands guided only by the sun, stars, wind, and waves. By the early 1970s, though, the Indigenous culture in places like Hawaii had been submerged and devalued by centuries of colonialism. Attitudes were shifting, and Finney's Polynesian Voyaging Society soon became a focal point for an emerging cultural renaissance that celebrated traditional language, music and dance, and voyaging traditions. But in 1973, when the society was founded, the hard truth was that no one in Hawaii knew how to get to Tahiti without a map and compass.

The same thing was true throughout Polynesia: the ancient navigational techniques, once carefully preserved through oral traditions passed on from generation to generation, had been forgotten once modern techniques arrived. There were a few scattered exceptions. In 1967, a New Zealand physician and adventurer named David Lewis arranged a highly unusual research fellowship with the Australian National University that involved sailing around the Pacific learning about traditional navigational techniques. On isolated islands in Polynesia, Micronesia, and Melanesia, he came across a few solitary navigators who still carried their ancestors' knowledge.

One such holdout was Pius Piailug, who had begun training with his grandfather in the 1930s on the tiny and isolated coral island of Satawal, in Micronesia, at the age of four or five. He was universally known as Mau, from the Satawalese word meaning "strong," because of his willingness to sail in bad weather. Navigational knowledge was traditionally considered sacred, never to be shared with outsiders. But Mau, in his early forties, was the youngest navigator left on Satawal and had no apparent heirs. Realizing his knowledge was in danger of being lost with him, Mau agreed to come to Hawaii and guide the *Hokule'a* to Tahiti, while explaining his methods to Lewis, who would also accompany the voyage.

No maps or navigational tools were allowed on the *Hokule'a*, and even the crew's wristwatches were stored in a locked box, since accurate timekeeping can be a navigational aid. That left Mau as the final arbiter of not only the route, but also official ship time. "Mau, midnight yet?" someone would inevitably ask on overcast nights, waiting eagerly for the end of their four-hour watch. Mau would squint into the featureless sky above him and, with a straight face, reply, "No, I think five minutes to."

These starless nights, as Andrew Sharp had warned, were navigationally challenging. Finney and his shipmates had to steer blind, maintaining a constant angle to the prevailing winds to keep moving in a straight line. That worked reasonably well in the reliable northeast trade winds above the equator and the similarly reliable southeast trade winds south of the equator. But getting from Hawaii to Tahiti required crossing the equator and sailing through the doldrums, a region of shifting and often nonexistent winds, complicated by unpredictable currents and frequent thunderstorms, that can be as much as three hundred miles wide. Thirteen days after leaving Honolulu, the *Hokule'a* sailed through a curtain of nighttime rain into the glassy waters of the doldrums and sputtered to a near halt.

On Captain Cook's first voyage, he and his crew were astonished to find that, during nearly a year of sailing, Tupaia could always point in the direction of his home island of Tahiti. As the *Hokule'a* meandered through the doldrums, Mau's sense of direction proved to be just as unshakable. On dark nights, the wind was no longer a reliable compass point. Finney took over the 4:00 a.m. watch one overcast night and found a steady breeze blowing, which gave him a directional cue to maintain the boat's southeast heading—but when the clouds cleared half an hour later, the moon was on the wrong side of the canoe. The wind had gradually shifted by 180 degrees, and the boat, blindly following the direction of the fickle breeze, had done a U-turn and was heading back toward Hawaii.

Mau's abilities weren't magic. In his discussions with traditional navigators across the Pacific, David Lewis wrote, "never once did anyone lay claim to any form of 'sixth sense.'" Instead, it was a matter of picking up scattered clues from a variety of sources: the wind, passing swells, the heavens, the pitch of the boat: "the most sensitive balance," a veteran skipper told Lewis, "was a man's testicles." Ten visible stars were enough to hold a steady course; five could do in a pinch, and sometimes even one or two stars, peeking through a gap in the clouds, was enough to confirm a heading.

All this required constant vigilance. "They say you can tell the experienced navigators by their bloodshot eyes," one anthropologist wrote. Mau seldom slept, and even when he did, the crew sometimes had to wake him up when they became disoriented, Finney later recalled: "Then Mau blinks his bloodshot eyes open, surveys the dark sea and overcast sky and renders his verdict: sometimes the hoped-for 'Okay,' but more often a deflating statement such as 'I think we go north. Turn around.'"

Not surprisingly, the stress of being stuck in the doldrums brought underlying tensions on the boat to a boil. Well before the trip started, a split had formed between those who, like Finney, saw the trip primarily as a scientific experiment, and the crew composed primarily of Native Hawaiians, who saw it as the spearhead of a cultural reawakening. To Finney, messages from the outside world, modern adaptations to the sailboat's rigging like adding a jib, and deviations from the traditional voyaging diet like contraband marijuana and beer were all no-nos. To the crew, few of whom had any experience with long-distance ocean voyaging, these restrictions seemed arbitrary and unnecessary, and after a week in the doldrums they were on the verge of mutiny.

Finally, on the twentieth day of the voyage, they passed through another curtain of rain and emerged from the doldrums into the southeast trade winds. They didn't leave all their problems behind them: an altercation later in the trip culminated with one of the disgruntled crew members punching Finney, Lewis, and the ship's captain, Kawika Kapahulehua, in the face. They also had no way of knowing how far to the west or east they might have drifted during their week in the shifting currents of the doldrums. But with the wind now reliably in their sails, they made steady progress until, on the fateful thirtieth day, Mau predicted that they would reach the Tuamotu Islands, about two hundred miles north of Tahiti, the next day. He was right. They stopped for a night of celebratory feasting on the island of Mataiva, then sailed on for the final leg, their bearings no longer in doubt.

The scene in Tahiti when they arrived is the stuff of legends. The governor had declared a national holiday, and more than fifteen thousand people crammed the harbor in Papeete, perched in trees, atop buildings, and knee-deep in the surf. During the month-long voyage, while Finney and his shipmates were off-grid, all of Polynesia had been transfixed, following the *Hokule'a*'s progress via location updates sent from a trailing escort boat. Despite the strife on board, the voyage had succeeded beyond anyone's wildest expectations in catalyzing renewed interest in Polynesian voyaging, and renewed pride in Polynesian culture. These days, the *Holuke'a* and other vessels like it sail on, including a three-year circumnavigation of the world completed in 2017, guided by a new generation of navigators who have learned the traditional ways.

From a scientific perspective, the *Hokule'a*'s voyage doesn't prove that ancient seafarers set out from southeast Asia to explore and colonize the islands of Polynesia. After all, Thor Heyerdahl rafted in the *Kon-Tiki* from Peru to the Tuamotus in 1947 to support his own pet theory that Polynesia was settled from South America—but despite the sensation his voyage caused, that theory has been long dismissed on the basis of archaeological and anthropological evidence. The

Hokule'a did prove that, despite the objections of Andrew Sharp and others, deliberate colonization was indeed *possible*. And in the half-century since then, a growing body of archaeological and genetic data has bolstered that view. The details and dates are still being revised, but the general thrust of the oral histories collected by Abraham Fornander and others seems to be correct: Polynesian islanders really were—and are—the Vikings of the Pacific.

Do humans have an innate desire to explore the unknown, something that differentiates us from maple seeds helicoptering in the breeze or lemurs rafting by happy accident to Madagascar? The settlement of Polynesia's imponderably far-flung outposts is as good a smoking gun as we've got. "There is something different in how our species, *Homo sapiens*, have been able to populate remote islands," Dylan Gaffney, an archaeologist at the University of Oxford who studies ancient water crossings and island settlements, told me.

Compared to other human species like Neanderthals, and perhaps even to other animals, it's probably a difference in degree rather than a totally new trait, Gaffney says. There's evidence—still hotly debated—for water crossings by earlier Stone Age humans, for example to Crete more than a hundred thousand years ago. These shorter trips don't require the same level of skill or, more crucially, show that the voyages were intentional. What distinguishes our species is the scale and ambition of our water crossings, and our ability to survive in new and challenging environments upon arrival. To Gaffney, it's not so much that humans were born to sail across the ocean, but that we have the "adaptive flexibility" to see and exploit the opportunities we encounter.

That combination of adaptive flexibility and ecological contingency—the luck of what's available where a given population happens to live—might explain why *Homo sapiens* made water crossings of hundreds of miles, far beyond what was visible on the horizon, and Neanderthals didn't. It might also explain why Polynesians supposedly have such an affinity for the sea, and for the unknown, even compared to other populations that migrated to various far-flung corners of the world. They grow up on beaches and boats, their ancestors came from distant islands, their nearest neighbors are across the water—of course they're eager to set sail. But there's another possibility, one that emerged from a controversial 1999 study and continues to reverberate: maybe it's in their genes.

2

CHASING DOPAMINE

It's not hard to understand why Timoteo Ruvalcaba migrated from Mexico to the United States. He'd left his hometown of Ameca, near Guadalajara, at the age of seventeen, to seek work in the cotton fields of Sonora. But that plan failed, and in November 1972 he decided to cross the border to join his brothers in California. It took him four months and four attempts to get across without being rounded up and sent back by the border patrol. Finally, a coyote smuggled him over the border in the spare tire compartment of his car to the Los Angeles suburb of Pacoima, where he spent his first year working in a factory assembling trailers.

In the decades that followed, Ruvalcaba moved from job to job, slowly climbing the ladder, accumulating savings, and eventually getting American citizenship. He launched a series of entrepreneurial ventures: importing clothing, selling groceries. By 1992, when he sat down for an oral history interview with researchers from the Mexican Migration Project, he was wistful about the life he'd left behind in Mexico but confident he'd made the right choice economically. "We all know that the United States is better than our homeland to make money," he said.

That's the classic narrative of modern migration: the push of a perceived lack of opportunities in one place, the pull of greener fields somewhere else. Timoteo Ruvalcaba's journey conforms to our expectations. But when he arrived in the United States, his older brother José was already there waiting for him—and José's journey is harder to explain. At the time José left Mexico in July 1972, he had a permanent job at a bank in Ameca, where he'd been working for eleven years. He already had a wife and five children, a stable life, a comfortable future. He didn't need to emigrate; but he was curious. He wondered what life would be like on the other side. "We jump borders," as Svante Pääbo said. "We push into new territory even when we have resources where we are." José took a leave of absence from his job and went up to Mexicali to check out the border. He figured he could make it across, so he quit his job and took the plunge.

José didn't have an easy time in California. He was the trailblazer, venturing where none of his friends or family had gone before. He started at a cannery in Watsonville, then began washing dishes at a restaurant in Hollister, moved to another cannery, became a housekeeper at a motel in South Lake Tahoe, a custodian at the Sahara casino, a busboy, a waiter, and on and on. He lived under the constant threat of deportation, using false documents under the name José Angel Ramírez for his first five years in the United States. Meanwhile he lined up jobs for his brothers, including Timoteo, brought his wife and children to live with him, and welcomed more and more of his extended family. Two decades after he arrived, he estimated that there were five thousand Latinos living in South Lake Tahoe—and 80 percent of them were from his hometown of Ameca.

Still, at the time of his oral history interview in 1992, he didn't own a house. He'd been in a position to buy one several times, but each time he'd asked himself: "Well, these twenty thousand dollars that I have saved, what can I do with them? Either I buy a house or I start a business." And he had always opted for the latter: he opened a restaurant, sold firewood, ran a tortilla factory. The path seldom ran smoothly, and at times he was on the verge of bankruptcy, but he kept on seeking his big break—"Because of the adventurer's spirit that I have, of always having to explore, to see whether there's more money here than over there."

The stupendously improbable settlement of Polynesia implies that humans must have some sort of compulsion to explore. No pattern of overpopulation or internal strife can explain how we reached all those islands; nor can any mathematical model of population diffusion. This argument is essentially cultural: in a particular place in the world, at a particular time in history, there was a society where exploring was highly valued. That doesn't mean that our current societies have the same characteristics. In fact, the near-disappearance of Polynesian navigational skills prior to the voyage of the *Hokule'a* could be interpreted as evidence that exploring isn't a universal and enduring aspect of being human.

But there's another line of evidence that connects the Polynesian experience to a broader view of who decides to explore and why. Whatever social and economic forces spur some Mexicans to migrate to the United States, they don't make *everyone* migrate. What explains why one person chooses to migrate, while his neighbor—or perhaps even his brother—chooses to stay? Those individual similarities and differences might, paradoxically, reveal something more universal, because patterns that run in families might be driven by genetic factors. And there's one particular gene, tied to how our brains process the neurotransmitter dopamine, that is a particularly compelling candidate.

In the mid 1990s, Chuansheng Chen was a young developmental psychologist at the University of California, Irvine who specialized in cross-cultural comparisons. One of his research projects focused on attention deficit hyperactivity disorder, or ADHD, a condition whose diagnosis was rapidly increasing at the time. Twin studies suggested ADHD was about 75 percent heritable, so a colleague of Chen's named James Swanson was searching for genes that might explain it.

Connecting specific genes to an ADHD diagnosis was far from straightforward, though, because the diagnosis itself remained an imprecise science. "It's based on reports from parents and teachers," Chen explains. "But that's based on how you stand out from the rest of your peers, right? If everybody is quiet and you speak one word, you're a 'talkative' person." Rates of ADHD in China appeared to be similar to those in the United States, but when Swanson traveled to China he saw that children were strikingly docile in class. By American standards, almost no one seemed to have ADHD.

The differences between Chinese and American schoolchildren

could have been cultural. But in the 1970s and 1980s, research with newborn babies had also begun to hint at inborn differences in temperament between cultures. Swanson's team had identified a genetic variant that affected a dopamine receptor in the brain called DRD4, and it seemed to be more prevalent in children with ADHD. Chen wondered whether the ADHD-linked version of this gene might be more common in some populations than others—and in 1996, geneticists at Yale published data that confirmed his hunch.

The Yale team had analyzed the DNA of more than 1,300 people from thirty-six different populations, and found three distinct versions of the DRD4 gene. Most notable from Chen's perspective was that the version associated with ADHD was virtually nonexistent in East Asian populations, while more than three-quarters of people in Indigenous South American populations had it. This was a hugely important result for the then-nascent field of genomics, which until then had taken most of its data from European subjects: it showed that you couldn't assume that what looked like "normal" DNA in one country would be the same in another country.

But what struck Chen was that DRD4 wasn't just any old gene. Swanson's team had noticed a connection between DRD4 and hyperactivity; other experiments had linked dopamine receptors to exploratory behavior in mice. And just a few months before the Yale results were published, in early 1996, a team in Israel had linked DRD4 to novelty seeking—the first time a stretch of DNA associated with brain function had been reliably linked to a normal personality trait. What if the geographical variation in DRD4 genes wasn't just a consequence of random mutations? What if, instead, that thirst for novelty was the *reason* for the variation?

Chen's idea was that the global distribution of novelty-seeking genes might be a consequence of how far various populations had migrated in the past. He pooled genetic data from 2,320 people in thirty-nine different populations: Sardinians, Ethiopian Jews, Mayans, and so on. Then he teamed up with an anthropologist to estimate how far each of these populations had migrated between 1,000 and 30,000 years ago from the original homeland of their language group.

He ended up with six migration routes. One was the movement of Native American groups from northeast Asia down the coast of North America and ultimately to the southern tip of South America: the Cheyenne had traveled 6,600 miles; the Mayans 8,600 miles; the Guahibo, in Colombia, 9,900 miles; the Karitiana, in Brazil, 11,200 miles. Other routes traced the spread of Indo-European groups across Europe; of Jews from Israel to Yemen, Ethiopia, and Italy; of African groups from the center of the continent to the southern tip.

The results, which Chen published in 1999, showed a clear pattern. Within each migration route, the proportion of the population with the novelty-linked version of the DRD4 gene increased by 4 or 5 percent for every 1,000 miles of migration. Sardinians, Yemeni Jews, Han Chinese, and the San in South Africa, all of whom had been living in roughly the same location for tens of thousands of years, had very low levels of the novelty gene. At the other extreme, the variant showed up in 78 percent of Colombia's Ticuna population, whose migration distance of 11,300 miles was the longest in the study. Nomadic societies tended to have a higher prevalence of the novelty gene than settled ones, independent of their original migration distance.

Here, then, was a smoking gun—not just for the narrow idea of an "explorer's gene," as the novelty-seeking variant of DRD4 soon came to be known, but for a broader view of how societies develop. "Before that point, no one believed that any human behaviors would have consequences on human evolution," Chen says. "We always assume that evolution is so slow." And there was a further telling detail: DNA analysis suggested that the novelty-seeking mutation in DRD4 first emerged in humans around between forty thousand and fifty thousand years ago—right around the time when our ancestors began their long, multipronged march to the farthest corners of the world. It was a march, the findings hinted, spurred in part by dopamine.

Thanks to bestselling books with titles like *Dopamine Nation* and *The Molecule of More*, dopamine has become a shorthand for pleasure, reward, and even addiction. Whether you're checking your phone at the dinner table or snorting lines of coke in the bathroom, it's all supposedly in pursuit of a dopamine rush. But its actual role in determining our behavior is more nuanced than its popular reputation as the "pleasure chemical" suggests.

The dopamine story starts in August 1957, when a British researcher named Katharine Montagu published her findings about an unidentified chemical that she dubbed X. Montagu had found traces of X in the brains of rabbits, guinea pigs, rats, chicks—and in a human brain. Its chemical properties seemed similar to a relatively obscure compound called hydroxytyramine, which was sometimes referred to by the nickname dopamine, because it could be synthesized from an amino acid called dopa. Dopamine wasn't thought to occur in brains, but a few months later a Swedish neuropharmacologist named Arvid Carlsson confirmed Montagu's findings. Carlsson went on to show that dopamine wasn't just an incidental player in the brain's workings: it was a neurotransmitter, responsible for relaying signals from one neuron to the next.

Carlsson's claims were controversial, because they landed him in the middle of a famous scientific debate known as "soups versus sparks." Since the beginning of the twentieth century, pharmacologists and physiologists had been arguing about whether neurons used chemical or electrical signals to communicate with each other. We now know that both sides of the debate were correct: electrical signals carry messages from one end of a neuron to the other, and chemical signals—neurotransmitters—carry messages across the small gap between one neuron and the next. But most scientists at the time remained skeptical about chemical transmission in the brain, and when Carlsson presented his results at a symposium in London in 1960 the response was almost unanimously negative. The conference host, in his closing remarks, insisted that nobody really believed that dopamine was involved in the function of the brain. "But this was what I had insisted upon throughout the meeting," Carlsson later recalled, "so the clear message to me was that I was nobody!"

Carlsson had the last laugh: the quote above is from his acceptance speech for the Nobel Prize in 2000. And it didn't take long for the powerful effects of dopamine to become clear. Carlsson found particularly high levels in regions of the brain associated with motor control, which suggested that it might be linked to the movement disorders associated with Parkinson's disease. In a 1961 paper describing the first experiments with L-dopa, a drug that boosts dopamine levels in the brain, researchers describe mute, bedridden Parkinson's patients suddenly regaining speech and rising from their beds to run and jump. L-dopa later played a similar, albeit temporary, role for the patients with sleeping sickness described in Oliver Sacks's 1973 book *Awakenings*.

The idea that dopamine might be linked to pleasure emerged in the 1970s. Scientists had discovered what appeared to be "pleasure centers" in the brain: if you wired up a rat so that it could administer a jolt of electricity to one of these pleasure centers, it would happily self-stimulate for as long as you let it, forgetting even to eat and drink. These pleasure centers turned out to coincide with regions of the brain where dopamine was produced; and if you administered a drug that blocked the effects of dopamine, the rat would lose interest in self-stimulating. The implication, as one scientist wrote in 1980, was that these dopamine centers were where "sensory inputs are translated into hedonic messages we experience as pleasure, euphoria, or 'yumminess.'"

But this simple picture didn't last long, thanks to two key experiments. The first, published by Kent Berridge and Terry Robinson of the University of Michigan in 1989, tested the pleasure theory by chemically wiping out dopamine neurons in the brains of rats. As predicted, the rats lost interest in food: left to their own devices, they no longer seemed to *want* even the sweetest offerings. But to Berridge and Robinson's surprise, when sugar was squeezed directly into the rats' mouths, they still *liked* it just as much as before, as revealed by their lip-licking and other facial expressions. Dopamine, it seemed, was the drug of motivation and desire rather than in-the-moment pleasure (which, Berridge says, more recent research suggests is mediated by other neurochemicals like endorphins and endocannabinoids, the body's internal versions of opioids and cannabis).

The second experiment, published in 1993 by a German-born neuroscientist then working in Switzerland named Wolfram Schultz, linked dopamine to the concept of reward prediction error. Schultz used electrodes to monitor the response of dopamine neurons in the brains of a pair of long-tailed macaques to the reward of a swig of apple juice. Initially, the juice generated a spike of dopamine activity. But if the experiment was repeated over and over, the dopamine response eventually faded away. And if they learned to expect juice but *didn't* receive it, their dopamine levels would actually drop below normal. Schultz suggested that, rather than signaling pleasure, dopamine's role was to signal the presence of a reward that was better than expected.

There's still no final consensus on exactly what dopamine's role in the brain is—or rather, what its many roles are. But both those experiments—Berridge and Robinson's wanting versus liking, and Schultz's reward prediction error—remain influential in our understanding of motivation and learning. And both have obvious connections to novelty-seeking and exploratory behavior. We explore because we want something else more than we like what we already have; dopamine is what drives that wanting. No matter how great the status quo is, it soon becomes predictable. To get the better-than-expected rewards that trigger our dopamine neurons, by definition, requires trying something new. Sure enough, if you artificially boost dopamine levels in the brain, monkeys ramp up their exploratory behavior.

So where does Chen's DRD4 receptor, which is seemingly linked to novelty-seeking and the history of human migrations, fit into this picture? It's one of at least five types of dopamine receptor that help ferry signals from one neuron to the next and determine how strongly or weakly those signals are transmitted. There's one particular part of the DRD4 gene where a sequence of forty-eight DNA units repeats somewhere between two and eleven times. The precise structure of the DRD4 receptors in your brain depends on how many of those repeats are encoded in your genes.

Until about forty thousand to fifty thousand years ago, around the time of the Great Human Expansion, most people seem to have had four repeats. Then, thanks to the random mutations that are constantly introducing subtle changes to our gene pool, a version with seven repeats emerged. This version of the receptor exhibits "a pronounced gain of function," scientists have found: you get a bigger spike of reward signaling in the brain in response to a positive prediction error, and a bigger drop in response to a negative one. As a result, this seven-repeat version (along with other variants like a two-repeat version that seem to have similar effects) primes those who have it to be extraresponsive to unexpected rewards. For some people, in other words, the dopamine-mediated thrill of discovering something new and unforeseen is more intense. Such people, decades of studies have found, tend to be more impulsive, more curious, more distractible, and perhaps just a little bit more likely to venture into the unknown.

The story of José and Timoteo Ruvalcaba comes from the Mexican Migration Project, a cooperative initiative headed by Jorge Durand, an anthropologist at the University of Guadalajara, and Douglas Massey, a sociologist at Princeton University. Starting in 1982, the MMP began sampling communities across Mexico to collect social and economic information about those who emigrated to the United States and those who stayed behind.

On a macro level, it's easy to understand the economic forces that drive migration. But what explains why one person decides to stay and another decides to go? "It is generally accepted by migration scholars that immigrants are positively self-selected with respect to characteristics such as drive, ambition, work motivation, health, and willingness to accept risks," Massey says. He and Durand have also used their database to make a convincing case for what's known as the network theory of migration. If you know other people who have migrated, you can get tips on how to cross the border, what types of jobs are available, where to find someone who speaks your language, and so on. All these elements of your social network alter the risk-reward profile of migration and thus influence who decides to go. That, in a nutshell, is how four thousand people from the Ruvalcabas' hometown of Ameca ended up in South Lake Tahoe: it's social network theory on steroids.

In 2011, Durand and Massey decided to make the MMP database freely available to other researchers. At that point, it contained information from 144,258 people in 134 communities (it has continued to expand since then), with richly detailed information on the social links between participants. And that's not all: implicit in the family trees sketched out for each household was information on genetic relatedness. Sure, I might be more likely to migrate if my sister has already migrated, because of social factors like the information she can provide and the generous checks she's sending home. But maybe there's also something in our genes—in our shared genetic inheritance—that predisposes both of us to migrate.

One of the researchers working with the newly opened MMP database was Jeffrey Napierala, then a graduate student in the University of Albany's Center for Social and Demographic Analysis. Napierala had read Chen's work on the DRD4 gene, along with subsequent studies that had replicated the effect, and wondered whether that gene or others like it might have left a mark in the MMP database. Napierala teamed up with Timothy Gage, the head of the Center for Social and Demographic Analysis. Gage is an academic hybrid: a demographer—someone who studies how human populations live and move and grow—who did his doctorate in anthropology but also did postdoctoral research in a genetics department in the early 1980s, when the field was young. "When I started working with sociologists, I didn't tell them I had a genetics background, because they would have run me out of the place immediately," he jokes. "Genetics has been slow to come into sociology."

Gage was intrigued by the heretical idea of a heritable influence on migration, but he wasn't optimistic that it would show up in the real-world messiness of the MMP's data. Still, he gave Napierala a green light to run the analysis as a class project. Napierala grouped pairs of subjects within a given household into five categories of relatedness: same-sex twins, opposite-sex twins, full siblings, half-siblings, and unrelated household members. The social ties and other environmental factors within a household should be roughly similar, but the genetics are different. For example, same-sex twin pairs, which might be either identical or fraternal, will share more DNA with each other than opposite-sex twin pairs, which are always fraternal.

Napierala's hunch paid off. "When he showed me the numbers," Gage recalls, "that's when I said, 'Wow.'" Pairs of same-sex twins were more likely to make the same decision about whether to migrate or stay home than pairs of opposite-sex twins. Similarly, the decisions of full siblings—like José and Timoteo Ruvalcaba—were more correlated than those of half-siblings or unrelated family members. The findings don't prove that your genes make you migrate, especially since we don't even know which versions of the DRD4 gene people in the database had, but they're suggestive.

Other species offer even stronger evidence of DRD4's influence. A long-term study of wild rhesus monkeys on an island off the coast of Puerto Rico found that juveniles with the long version of the DRD4 gene—the one associated with novelty in humans—scored higher on measures of behavioral restlessness and ventured away from their mothers more often. A similar study of yellow-crowned bishops, a songbird from Africa that has recently expanded its range into Spain and Portugal, found that variations in the DRD4 gene predicted their level of interest in unfamiliar objects such as batteries and slices of apple.

Compared to the messy complexity of human decision-making, animal data also offers a clearer look at how these sorts of behavioral differences can influence migration. In one study, researchers compared the frog populations in mainland Sweden with those on small offshore islands colonized by frogs within the past few hundred years. Eggs from both places were hatched in the laboratory, so that all the frogs grew up in the same conditions. In tests of exploratory behavior, tadpoles and froglets from island eggs showed dramatic differences: they were quicker to emerge from a safe hiding spot into an unfamiliar environment, and explored as much as 50 percent more territory after emerging. The most likely explanation is that only the boldest and most curious frogs were likely to make the deep-water crossing to the islands, which are now populated by descendants who have inherited their wanderlust.

In some cases, you can see this process playing out in real time. House sparrows were introduced to Mombasa, on the coast of Kenya, around 1950, and have been spreading inland ever since. In behavioral tests on sparrows from cities across Kenya, exploratory behavior increased in direct proportion to how far from Mombasa the sparrows lived. The most intrepid sparrows were those at the leading edge of the expanding range, which is now more than five hundred miles into the highlands near the Ugandan border.

The same is true for cane toads in Australia, introduced—in a colossal act of misjudgment—in 1935 to control beetles in sugar plantations. Since then, they've been hopping west across the Australian interior, growing to a population of more than two hundred million and wreaking havoc on local ecosystems. In the initial decades after their release, their range spread westward by about six miles each year. But the rate of progress has steadily accelerated, and by the early 2000s they were overrunning nearly forty miles of new territory per year. Once again, controlled studies of exploratory behavior—how long it takes to emerge from a hiding place into an unfamiliar environment, how much of that environment they explore in a given amount of time—show that toads at the western edge of the expanding range are significantly more eager to explore than those in places that were overrun decades ago.

The cane toad data also punctures any lingering illusion that exploratory behavior is a simple matter of which DRD4 gene you've got. The toads in the vanguard of expansion differ in a variety of ways that extend beyond novelty-seeking. They're more likely to travel in straight lines; they spend less energy on their immune systems; and they even have longer legs, enabling them to cover more ground every night. It makes sense that long-legged, fast-moving toads are the first to reach new territory, but the changes run deeper than that. The long-legged toads at the front of the wave then mate with each other, producing even longer-legged toads: current toads have legs that are as much as 10 percent longer than the ones that first arrived in Australia in the 1930s. In times of change, evolution can move very, very quickly—by leaps and bounds, even.

It's a lot easier to measure toad legs than it is to quantify the itch to explore. But the toad findings, along with the animal data more generally, highlight some of the unresolved questions about what Chuansheng Chen's DRD4 data really tells us about human migration. One view is that those with the right version of DRD4, or whatever other traits might predispose someone to explore, were more likely to migrate in the first place. Another view is that everyone was equally likely to migrate, but those with novelty-loving traits were more likely to prosper—and thus pass on their genes—in unfamiliar new environments. The cane toads seem to take a middle path between these two extremes. A diverse mix of characteristics like curiosity and long legs sends some toads to the front of the expanding range. Then, once they're there, they pass those traits on to their offspring so that they become more and more common among the leading wave of toads—a phenomenon biologists call gene surfing.

Among humans, it's clear that some people are more predisposed to emigrate than others. That's presumably why, for example, the 1.5 million Scandinavians who emigrated to the United States and elsewhere in the late nineteenth and early twentieth centuries tended to have less common first names than those who stayed behind: unusual first names are a reliable proxy for individualistic streaks within a family. (It's also no coincidence, some researchers believe, that the collectivists who stayed behind have ended up building some of the strongest social welfare states in the world.) But there's also evidence that gene surfing occurs as readily in human populations as it does in cane toads, thanks to patterns of reproduction among those who have already emigrated.

An unusually well-documented example of gene surfing comes from the waves of French-Canadian settlers who began spreading into the forests northeast of Quebec City along the Saguenay River starting in the 1830s. Detailed parish registers dating back to the seventeenth century allowed researchers to trace the family trees of those in the region. As new settlements were hacked out of the bush farther and farther upriver, restless young opportunity-seekers led the way— and were rewarded for their gamble. Those at the front tended to get married earlier and have more children—9.1 versus 7.9—compared to those downriver. And like the cane toads, their reproductive edge was passed from generation to generation: those at the wave front during those crucial decades of expansion passed on roughly twice as many genes to modern *Québécois* as those lingering in more settled places. "Exploration may thus create a self-reinforcing loop," the science journalist David Dobbs argues, "amplifying and spreading the genes and traits that drive it."

The conclusion from all this genes-and-migration data, you might think, is that only some people are wired to explore. But recent advances in how scientists study ancient migration patterns have changed that picture. In 2015, a research team led by Ron Pinhasi, then at University College Dublin, located the ultimate mother lode of ancient DNA. The petrous bone, located at the base of the skull and housing the hearing organs of the inner ear, is among the densest and hardest bones in the human body. Its density means that it often survives even after other skeletal remains have crumbled to dust—and, more crucially, that DNA within it is preserved even in hot and humid environments where organic material usually breaks down quickly. Grinding up the inner ear portion of ancient petrous bones yields about a hundred times more salvageable DNA than other bones, and this insight has helped open a radically new window onto human population movements in the distant past.

The key insight from these studies, according to Harvard University geneticist David Reich, is that we've been on the move for most of human history. "Populations today almost never descend directly from the populations that existed in the same place even ten thousand years ago," he says. In its original form, the Out of Africa hypothesis imagined human populations spreading steadily outward from their common origin, like the expanding stain of a spilled wine glass, until they covered the globe. But the actual course of history has been much less linear, with populations moving back and forth within and across continents and often mixing along the way.

The British Isles, for example, were overrun about 4,500 years ago by people whose ancestry can be traced all the way to the Central Asian steppe, mixing with previous Neolithic farmers and Mesolithic hunter-gatherers and subsequent waves of Anglo-Saxons, Vikings, and others. In Africa too the populations who stayed behind after modern humans began expanding into Asia and beyond weren't frozen in amber. Instead, various groups crisscrossed the continent. The Bantu culture, for example, originated in present-day Nigeria and Cameroon then, around four thousand years ago, began a long march all the way to the southern tip of Africa, five thousand miles away. This revised understanding of our historical wanderings suggests that virtually all human populations have probably been shaped by their travels, with natural selection favoring traits that encourage or facilitate exploration.

There's also reason to believe that the urge to explore will vary considerably within any given population, thanks to what geneticists call "frequency dependent selection." Some traits, like human height, seem to converge over generations to a fairly narrow range around an optimal value. Other traits, however, don't have a unique evolutionarily "correct" answer. For example, "there may not be one optimal personality," explains Luke Matthews, an anthropologist who replicated and extended Chen's findings on DRD4 and migration in a 2011 paper. A society with a mix of introverts and extroverts might outperform both an all-introvert society and an all-extrovert one, for example. Same with aggression and cooperation, where the pros and cons depend on the frequency of those traits in the people around you, Matthews says. "It's most advantageous to be aggressive when no one is going to aggress back at you, right?"

A similar balancing act plays out with the various versions of the DRD4 gene. A 2008 study led by Dan Eisenberg, an anthropologist now at the University of Washington, tested the genes of pastoralists from the Ariaal tribe in northern Kenya. The Ariaal are traditionally nomadic, relying on herds of camels, cattle, and other animals for their sustenance. In recent decades, some have settled in villages and taken up agriculture. Eisenberg found that about a fifth of Ariaal men in both groups had the novelty-seeking version of the DRD4 gene.

Among the nomads, these men were better nourished, with higher body-mass index thanks to greater muscle mass. Among the villagers, it was the opposite: novelty-seekers were *less* well nourished.

It's easy to imagine how this pattern occurs. Restless explorers thrive in the constantly changing environment of nomadic life, but struggle in the repetitive monotony of settled agriculture. This is the same argument advanced to explain the evolutionary benefits of ADHD, which is also linked to the DRD4 gene, among our hunter-gatherer ancestors. But despite the apparent advantages, the novelty-seeking version of the gene never fully took over in any populations. The Ariaal results suggest that there are advantages and disadvantages to each version of the gene, depending on the context and environmental conditions. In any given population, you always want some people pushing back the borders of the map and others keeping the home fires burning.

The simple idea of an "explorer's gene," then, turns out to be not so simple. The grand historical movements of human populations do seem to have shaped us to thirst for new horizons and thrive when we get there. The fact that the novelty-seeking version of the DRD4 gene first arose around forty-five thousand years ago, just when humans began conquering the globe, is highly suggestive. But DRD4 is just part of the story. Complex behaviors like exploring are inevitably driven by an equally complex mix of genetic and environmental factors. After all, what differentiates us from archaic humans, or modern primates, or other animals more generally, isn't a single gene. It's a whole suite of abilities—call it behavioral modernity—along with the gradual accumulation of knowledge across generations. "It's biocultural," Matthews says. "We've got the boats and the Gore-Tex at this point."

What's clear is that our nomadic history has left an imprint on us. We all carry the legacy of genes that surfed on the breaking wave of human migrations—and some of us inevitably have a bigger helping of those genes than others. Matthews, for reasons that he still struggles to articulate, decided to leave his girlfriend behind and head into the jungle to study capuchin monkeys when he was a graduate student: "I'm just a regular guy who was raised in the suburbs of Philadelphia, right? But I went and lived in the Amazon for a year, without even phone access to the outside world." Now he's comfortably settled in Boston, working as a behavioral and social scientist at the RAND Corporation. He's married—his girlfriend waited for him—and has three kids. One of their family interests is following the latest news about space exploration. They don't all see it the same way, though. "My wife and my oldest are both like, no, they would definitely not be part of a one-way trip to a Mars colony," he says. "Whereas my daughter and I are like, 'Sign me up tomorrow! Launch me! Launch me on the one-way trip!'"

3

THE FREE ENERGY PRINCIPLE

Two paths diverged in a muddy wood, and the first team to reach the junction zigged right. Glancing quickly at my map without pausing to break stride, I zagged left along with my teammates Adrian and Nevin. The first checkpoint was at the top of a tree-lined ridge in the distance, and our route looked like the quickest and most direct approach. As we pushed on, Adrian glanced back over his shoulder at the stream of runners reaching the crossroads behind us. Every one of them was turning right. "Ummm," he gasped between breaths, "should we be worried about this?"

We were barely ten minutes into a race called Raid the Hammer, which was—in theory—about fifteen miles long. But there was no actual course. Instead, there were thirty-seven digital checkpoints stashed among the dense forests, vertiginous cliffs, and shoe-sucking swamps of a three-thousand-acre conservation area near Toronto. We had to hit them all, in the right order, using only a compass and the maps we were given just before the start. The organizers had said it should take between three and seven hours—depending as much on our route choices as our footspeed.

This was my first orienteering race, back in 2011. I was in my midthirties at the time and beginning to confront a puzzle that every longtime runner eventually encounters: When you've already run hundreds of races, what are you seeking in the next one? My motivations for running had stayed relatively constant as I progressed from eager middle-schooler to hopeful would-be-elite to wily veteran. The underlying theme was the desire to explore the uncharted boundaries of my abilities. How fast could I go? How much training could I handle? How deep could I dig? I knew that my answers to those particular questions were getting close to final. For some of my friends, that realization had been a signal that they were done with the sport. But I figured there might be other questions that running could help me answer.

In a way, every run is a voyage of discovery. How deep you can dig *today* is a question that can be answered only by doing it, regardless of how fast you've run in the past. There are equations that calculate your marathon time based on lab-measured physiological parameters like VO_2max, running economy, and lactate threshold; the essential point about running is that these equations are not very good. The conclusion is never foregone. "If you want to be a champion," the great American miler Marty Liquori once wrote, "you will have to win every race in your mind a hundred times before you win it in real life that last time." The corollary, though, is that all these imaginary wins mean nothing until you do it in real life.

Still, with enough repetition comes a degree of predictability. There are various ways that experienced runners maintain the spark of discovery: trying to beat their best age-graded time, tackling new distances or terrains, finding new running routes, traveling to distant races, and so on. I hoped to rekindle that spark at Raid the Hammer, the uncharted course a metaphor for my venture into an unfamiliar type of race. If you do an out-and-back route in a place where you haven't run before, the outbound leg always seems to take far longer than the return leg. This is because your perception of time depends in part on how much information is flowing into your brain. When you're navigating through new territory, your senses are heightened and you're taking in all the details of your surroundings. On the way back to your starting point, you know where you're going and can shift back toward autopilot.

At Raid the Hammer, the whole race would be run in that state of heightened awareness—or we would pay the price. A few minutes after the first intersection, we emerged from the forest to find a vast field of thistles, neck-high and densely intertwined, blocking our path.

It was now obvious why most of the other runners had turned right instead of following our "shortcut" to the left. By the time we fought our way through, we'd lost some blood, and almost seven minutes to the leaders. It was a classic rookie error. "Compared to a more experienced orienteer, you can hear a new cadet just smashing through things," Colonel Mike Hendricks, the head of the U.S. Military Academy at West Point's orienteering program, later told me. As the race progressed, the terrain got hillier, the forests got thicker, and the trails we'd been following gradually disappeared. Soon we were fighting through undergrowth, hopping over fallen trees, and sliding down ravine slopes. Every step was unpredictable—and that, it turns out, is precisely what my brain was seeking.

Human migration patterns imply that we seek out the unknown. Genetic data suggests that this impulse is wired into our behavior. But to understand *why* we're wired this way, we need to wade into the ongoing struggle to understand the most complex and enigmatic organ of all: the brain. Over the past two decades, a new way of understanding the brain has gradually taken hold among neuroscientists, cognitive scientists, psychiatrists, psychologists, philosophers, and others. It goes by various names and exists in multiple competing versions, reflecting its status as a theory-in-progress, but the general term I'll use here is "predictive processing," a phrase popularized by the British philosopher and cognitive scientist Andy Clark. In its most ambitious form, predictive processing offers explanations for how we understand the flow of information from our senses, what goes wrong in psychiatric conditions like depression and bipolar disorder, and what fundamental imperative all living things must obey. It also explains, at the most fundamental level possible, why we explore.

The basic idea of predictive processing is that perception is a two-way street. The traditional view is that we take in a bunch of "bottom-up" information through our senses, and send that data to our brain, which interprets it and determines what we're seeing, hearing, smelling, and so on. Then the brain decides what to do, and sends "top-down" signals to our muscles to carry out its instructions. That's how we've always tried to build robots, and it's how we've assumed humans are wired too. But if you look at how our nervous system is actually wired, you see a puzzling phenomenon: not only do signals travel from sensory organs like the eye up to the brain; they also travel simultaneously in the opposite direction, from the brain toward the eye. In fact, there are *more* connections from the brain to the senses than from the senses to the brain—"a strange architecture about which we are nearly clueless," as Patrick Winston, the former director of MIT's Artificial Intelligence Laboratory, once put it.

What's traveling along those reverse connections, according to the predictive processing view, are predictions. The brain is constantly formulating forecasts about what it expects to see and hear and smell, and comparing those top-down forecasts to the bottom-up data collected by its senses. When there are prediction errors, in which the incoming sensory data deviates from outgoing predictions, the brain updates its view of the world. When the prediction errors are sufficiently small, the brain simply assumes that its predictions are reality—meaning, Winston points out, that vision is "a matter of guided hallucination." Either way, the brain's overriding imperative is to minimize its own prediction errors.

The idea that the brain is a prediction engine has a long history: it's often traced back to Hermann von Helmholtz, the German physicist and physician who in the 1860s proposed that phenomena like optical illusions could be explained by that fact that our sensory perceptions are generated in part by the brain rather than by just the senses. Starting with a baseline prediction and then paying attention only to deviations from that prediction also turns out to be a very efficient way of running a brain. That's why information theorists use a similar approach for transmitting data and in compression protocols like JPEG files. A key breakthrough came in 1999, when Rajesh Rao and Dana Ballard presented a model of vision using what they called predictive coding. They were able to resolve some previously unexplained quirks in how we see the world by assuming that signals travel in both directions: predictions from the brain toward the eye, predictions errors from the eye toward the brain.

So far all we've explained is how we make sense of our senses. That's not nothing, Andy Clark points out: it demystifies the phantom buzz of a nonexistent cell phone in your pocket, explains why the lyrics of familiar tunes are easy to decipher even on a radio with poor reception, and reveals how millions of people in 2015 couldn't agree on whether a viral snapshot of an unremarkable dress was gold and white or blue and black. Which version you saw depended on what assumptions your brain was making about the lighting in the room. Morning people, who get up early and see more bright natural light during their waking hours, tended to see the dress as gold and white, it turns out; night owls, who have different assumptions about "typical" lighting levels, saw it as blue and black.

But perception is powerless without action. In 2005, a neuroscientist at University College London named Karl Friston published an attempt to explain a new idea he called the free energy principle, a mathematically rigorous version of predictive processing that promised to unite perception and action as flip sides of the brain's attempts to predict the world. Friston was already an enormously influential and famously inscrutable figure in brain science. His early work on brain imaging techniques radically transformed the field and has made him, by several different measures, the most influential and highly cited neuroscientist ever. As of 2024, according to Google Scholar, his papers had been cited more than 350,000 times by other researchers. But the free energy principle was a radical departure from his earlier work—a staggeringly ambitious attempt to fulfill a dream, nourished since he was a child, of "understanding everything by starting from nothing."

At its simplest, Friston's free energy principle states that all living things act to minimize surprise. Surprise here is a technical term. (It's "the negative log-probability of an outcome," Friston explains in one of his early papers.) But its meaning basically corresponds to our intuition. ("An improbable outcome is therefore surprising," he adds.) Whether you're a human or an amoeba, in order for you to exist there has to be a boundary between you and everything else in the universe. To maintain that boundary—to avoid simply dissipating into a shapeless collection of particles like a drop of food coloring in a bathtub—you have to resist the natural tendency for increasing disorder, or entropy. It turns out that minimizing surprise is equivalent to minimizing entropy, which in turn is equivalent to minimizing another mathematical quantity (borrowed from physics) called free energy.

The free energy principle, then, is a mathematical statement of what is required to exist as a discrete entity in the universe rather than being scattered to the winds. A perfect match between predictions and observations minimizes surprise—and free energy—and thus augurs well for your continued existence. When there's a mismatch, you have two choices. You can update your model: *I thought that tea was cool enough to drink, but the scalding sensation on my tongue tells me that it's not.* Or you can take action to update the world: *I'm going to blow on it, or add some milk, so that it will be cool enough to drink.* In this way, the goal of minimizing surprise explains both perception and action. We act, Friston says, in order to ensure that our predictions become self-fulfilling prophecies.

If this all seems a little arcane—well, you're not alone. Even scientists in the field struggle to make sense of it. Peter Freed, a psychiatry researcher at Columbia University, once described a journal club meeting where he and his colleagues grappled with one of Friston's papers: "There was a lot of mathematical knowledge in the room: three statisticians, two physicists, a physical chemist, a nuclear physicist, and a large group of neuroimagers—but apparently we didn't have what it took." Still, interest in the free energy principle, and in predictive processing more broadly, has continued to grow, because they generate all-encompassing predictions about how the brain should operate. Friston and others have used the framework to suggest explanations for how the brain is structured, what goes wrong in psychiatric disorders such as depression, and why we enjoy horror movies. Beyond the brain, they've also proposed free energy-based explanations for, among other phenomena, the growth of cancer cells.

There's one problem with predictive processing, though, at least for our purposes. It's what has become known as the Dark Room problem. If the prime directive of all living things is to minimize surprise, then a very effective strategy would be to lock yourself in a dark room and stay there . . . forever. At any given moment, you would be able to predict with exquisite precision what's going to happen next: nothing. Forget Polynesia; the free energy principle seems to suggest that we're not even wired to explore the room next door.

Friston's initial response to this critique was to note that natural selection and experience have equipped every life-form with very strong ideas about what organisms like themselves should expect in the world. We would find it very surprising, in the free energy sense of the word, to be attempting to breathe underwater. A fish would find it equally surprising *not* to be breathing underwater. Equivalently, sitting in a dark room getting progressively hungrier and more dehydrated would strike us as a very surprising situation to encounter, even though the moment-by-moment experience was highly predictable. Eventually, we would venture out in search of conditions that better match our ideas of how a person should be.

This "ho-hum human evolutionary story," as University of Toronto philosopher and cognitive scientist Mark Miller puts it, is the simplest way of wriggling out of the Dark Room problem—but, he adds, it's not a very satisfying explanation. It might explain why I don't lock myself in a closet, but it doesn't account for why I was drawn to try orienteering in my midthirties instead of running the same races on the same courses year after year for the rest of my life. If anything, evolution would consider the spectacle of a middle-aged man pushing himself to exhaustion for fun to be highly surprising—and thus a violation of the free energy principle—no matter how familiar or unfamiliar the racecourse was.

As Friston and others fleshed out the details of the predictive processing framework, a more nuanced and mathematically rigorous answer to the Dark Room problem emerged. The key is to broaden the definition of surprise. Instead of just minimizing my surprise *right now*, the revised free energy principle seeks to also minimize how much surprise I expect to have in the future. The best way to avoid future surprises is to gather as much information as you can about the world. And, indeed, if you work your way through the free energy equations that give Peter Freed and everyone else so many headaches, you find that calculating expected future surprise gives you a term that depends on how much present uncertainty a given action is likely to resolve. "Once you have those dynamics in play," Miller says, "you have a naturally curious organism that doesn't *want* to get stuck in a dark room." Minimizing future surprise, in other words, involves choosing the actions that offer the greatest potential to learn something new about the world—like opening the door and heading outside.

That's the theory. But does it really improve your ability to, say, blow the heads off bug-headed alien monsters with your phaser rifle? That's the question a team led by Rosalyn Moran, a professor of computational neuroscience at King's College London, tackled in a 2018 study. Moran and her colleagues programmed a pair of bots to tackle a modified version of *Doom*, the classic first-person-shooter video game.

One of the bots was programmed with standard machine-learning techniques. It could choose from three possible moves: pivot left, pivot right, or fire. It began with no particular preference for any move, and no knowledge of what each of these choices accomplished. But if its sequence of choices led it to kill a monster—to choose to fire when it happened to be facing directly toward a monster—it received a reward of a hundred points. By playing the game over and over, adjusting its parameters whenever it earned a reward, it gradually became a competent alien-slayer.

The other bot, instead of maximizing reward, sought to minimize prediction error as dictated by the free energy principle. It started with the prediction that it had a monster directly in its sights and was firing at it, but it too had no knowledge of what the possible commands—left, right, or fire—accomplished. Whenever a random sequence of moves led it to shoot a monster, it was "rewarded" by fulfilling its own prediction, reducing its prediction error to zero. In this sense, it was just like the machine-learning bot. But the free energy bot didn't just repeat the same moves over and over again. It was programmed to minimize future surprise as well as present surprise, which requires seeking out new prediction errors to resolve. The push and pull between these two goals produced a mix of behaviors: it repeated moves that led to shooting monsters but also deliberately tried new strategies that offered greater sources of uncertainty.

Initially, the simple reward-maximizing bot outperformed the free energy bot. But by the time they'd played the game a dozen times, the tables had turned. The reward-maximizer had gotten lucky a few times and kept trying to repeat its successful moves. The free energy bot, in contrast, had experimented with various different strategies, and pieced together an internal model of how the game—its universe—worked. No matter where the monster started, the free energy bot could figure out the appropriate sequence of moves to shoot it. After 128 iterations of the game, it left the reward-maximizer in the dust and was sniping as quickly and accurately as human players. According to Moran, the key difference is that the free energy bot, following its imperative to minimize present and future surprise, explored.

The Dark Room problem, as we've posed it, is one of those thought experiments that philosophers are famous for. Scientists who study predictive processing haven't actually locked people into dark rooms to see how they react. But other researchers have, at least metaphorically, and the results are both surprising and enlightening.

Most famously, University of Virginia social psychology professor Timothy Wilson and his colleagues asked volunteers to sit quietly in a room, without phones or other distractions, for periods of up to fifteen minutes. Most of the subjects reported that the experience was unpleasant, so Wilson upped the stakes: he repeated the experiment, this time attaching a silver chloride electrode to the subjects' ankles so that they had the option of self-administering mild shocks by pressing a button. Remarkably, 67 percent of the men and 25 percent of the women chose to zap themselves rather than sit quietly. Many of them administered several shocks, and one person did it 190 times. Wilson's results aren't a fluke: in addition to testing college students, he recruited volunteers ranging in age from eighteen to seventy-seven from the local community and found similar responses. In one follow-up study with 254 participants in Germany, almost 90 percent pressed the button. As bad as being shocked is, doing nothing is apparently worse.

Plenty of theories have been advanced to explain these seemingly irrational results: thinking is so hard that we take any distraction to avoid it; smartphones have destroyed our powers of concentration; we're no longer used to being bored. Wilson's explanation is that "the mind is designed to engage with the world," and the predictive processing theory more or less agrees with that take. But a 2016 follow-up study by two marketing researchers, Christopher Hsee and Bowen Ruan, takes the results one step further. It's not just that we're repelled by doing nothing. We're also attracted to the shocks—in exactly the way that the free energy principle predicts.

In Hsee and Ruan's study, subjects again had to kill time alone in a room. On a table in front of them were trick pens that could deliver an electric shock. Some were labeled red to indicate that the shocker's battery worked; others were labeled green to indicate that they didn't work; a third group was labeled yellow, meaning that it was unknown whether they worked or not. If, as Wilson argued, people just want to avoid doing nothing, you'd expect them to play with the nonshocking pens. If they're closet masochists or sensation seekers, you'd expect them to play with the shocking pens. Both those things happened—but the most popular option by far was the yellow-labeled mystery pens, which subjects clicked three times more often than either of the other choices.

The experiment reminds me of a memorable meal I had with friends at a restaurant in Sydney. We ordered a round of sushi roulette: six pieces of salmon maki, all identical except that one had been hollowed out and crammed full of wasabi. On cue, we all popped the rolls into our mouths, and one unfortunate soul—not me, thank goodness—ended up writhing on the floor and popping Tylenol for the rest of the evening. Again, if we wanted salmon maki, we could have ordered plain old salmon maki. If we wanted to obliterate our sinuses, there was plenty of free wasabi already sitting on our table. But what we really wanted—the reason I remember the meal fondly, and have been searching for sushi roulette on menus ever since—is that feeling of uncertainty followed by the moment of resolution.

As marketing profs, Hsee and Ruan are intrigued by our enjoyment of what they call the "teasing effect"—and they think advertisers should leverage it. David Ogilvy, the so-called father of advertising, famously advocated using the name of the brand within the first ten seconds of any ad. And if you ask people to choose, they'll generally opt to resolve uncertainty as quickly as possible. But if you actually measure enjoyment, it turns out that, contrary to their own expectations, people generally like a frisson of uncertainty: Will this pen shock me? Is this roll stuffed with wasabi? What brand is this ad selling, anyway?

Scientists who study curiosity and novelty-seeking have long noted that it's not just about gaining information. "Sometimes knowledge is sought with some immediate objective in mind," write George Loewenstein, a behavioral scientist at Carnegie Mellon University, and his colleagues, "but this makes up, on the whole, a surprisingly small part of our intellectual life." Instead, Loewenstein sees curiosity as a drive analogous to hunger or sexual desire. We've evolved these drives to fulfill useful ends: staying nourished or passing on our genes in the latter cases, obtaining potentially useful information in the former case. But in our day-to-day lives, these drives are decoupled from their original purpose. Sometimes we eat empty calories, have sex while using birth control, or read mystery novels.

The original formulation of the free energy principle didn't have much to say about why exploration might feel good. The drive to reduce prediction error might explain why you don't just sit in a dark room, but it doesn't explain why we sometimes actively seek out—or even construct—sources of uncertainty like a booby-trapped sushi roll. In a 2017 paper, Mark Miller and his colleagues Julian Kiverstein and Erik Rietveld proposed that we're motivated by what they call "the feeling of grip": the perception of how quickly we're reducing uncertainty, rather than the absolute level of uncertainty. "We're not interested in having *no* error," Miller says. Instead, we're looking for what he calls "digestible errors"—gaps in our knowledge that we can efficiently fill. The proposal builds on previous research linking emotional states to changes in the accuracy of our predictions: if we're gaining information faster than expected, we feel good; if we're learning more slowly than expected, we feel bad. When someone asks you how things are going, Miller and his colleagues argue, your answer (if you're being honest rather than just making formulaic small talk) reflects your general sense of how rapidly your prediction error is decreasing.

Exploring, in this picture, is what Andy Clark calls "slope-chasing." Imagine a chart that plots your current level of overall prediction error as time passes. In the hypothetical dark room, the graph will simply be a horizontal line, neither increasing nor decreasing. Once you start exploring out in the real world, the line will sometimes move upward when you encounter surprising situations, and then move back downward as you discern new patterns and incorporate this fresh information into your model of the world. It's these downward slopes, where our prediction error levels are going from high to low, that make us feel good—the steeper the better. That, in a nutshell, is why we seek uncertainty rather than avoiding it: you can't get the pleasure of reducing prediction error unless you have some to start with.

In fact, we're so eager for this feeling that instead of just chasing slopes of decreasing prediction error, we also construct them ourselves. Play, according to a theory advanced by Danish cognitive scientist Marc Malmdorf Andersen (along with Miller, Kiverstein, and Andreas Roepstorff), is a form of "slope-building" in which you deliberately construct the rules of your game to create new uncertainty and then resolve it. Does my oatmeal stick to the wall? Can I finish a marathon? Will my piece of sushi be the one filled with wasabi? In this way, the free energy principle not only gets us out of the Dark Room, but also explains why we actively seek out—and enjoy—the unknown.

This slope-chasing framework has an important corollary: if we're attuned to the feeling of *reducing* prediction errors, rather than simply liking accurate predictions, then we'll tend to seek out intermediate levels of uncertainty. Sitting in a dark room is uninteresting to us because there's no uncertainty to resolve, regardless of how predictable it is. But very high levels of uncertainty are also uninteresting, because you can only get the pleasure of reducing uncertainty if you can make sense of the situation. Trying to follow the plot of an artsy foreign-language film with no subtitles is a rich source of prediction error, but it's not *digestible* error. Somewhere between boringly easy and frustratingly incomprehensible is a satisfying midrange of uncertainty that you can grapple with productively.

The idea of a sweet spot is an important one that crops up all over the place in psychology. It's sometimes called the Wundt Curve, for

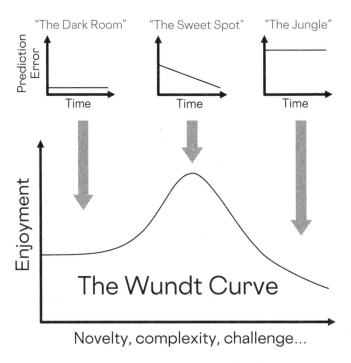

The Wundt Curve illustrates a general relationship between the novelty, complexity, or challenge of an experience and how much we enjoy it. In the Dark Room (left), the situation is too simple or familiar, so there is no prediction error to reduce. In the Jungle (right), the situation is so complex or unfamiliar that we can't make sense of it, so prediction error remains high. In the Sweet Spot (center), an intermediate level of novelty and complexity enables us to steadily reduce prediction error. It's this descending slope that we seek when we're exploring (slope-chasing) or playing (slope-building).

pioneering nineteenth-century German psychologist Wilhelm Wundt, who posited an "upside-down U" relationship between stimulus and subjective response: the more intense the stimulus, the more you like it . . . until, beyond a certain point, you start liking it less. The Wundt Curve is invoked to explain why we like art and music that's complex but not too complex, why video games are most engaging if they deliver an "optimal challenge" that's neither too easy nor too hard, and why flow states require a task that's just within your capabilities. It also explains why tastes change: those who listen to a lot of music get better at predicting what comes next, and as a result end up preferring more complex sounds.

This pattern is evident even in babies, who don't like being ignored but also (I've learned from experience) don't like excessively unpredictable stimuli like jack-in-the-boxes. Parents learn to modulate the level of surprise in games like peekaboo: playing tricks like making one adult disappear and another reappear in her place leads to fewer smiles. One study found that even seven- and eight-month-olds exhibit a "Goldilocks effect" in their thirst for unpredictability. Celeste Kidd, then at the University of Rochester, showed the infants a screen on which a sequence of various objects of interest—fire trucks, milk bottles, and so on—appeared from behind colorful boxes and then disappeared again. She measured their interest by tracking when their gaze wandered from the screen. When the sequence was highly predictable, the infants got bored quickly. When it was highly *unpredictable*, they also got bored quickly. When its predictability was somewhere in the middle, that's what kept their gaze the longest.

In a sense, the Wundt Curve is what makes exploring truly universal. As fascinated as I am by the DRD4 gene and other evidence for our intrinsic drive to seek the unknown, I have plenty of friends who say, "Well, that's just some people, not everyone. I'm not much of an explorer myself." What they mean is that they have no desire to parasail to the South Pole; or that they really do like ordering the same dish in the same restaurant every time they go out for dinner, thank you very much; or that they're simply too old or too busy for it. But that doesn't mean they're not exploring in other ways, as dictated by the shape of their personal Wundt Curve at that particular moment in their lives. "That sweet spot is a moving target," Miller says. "Because as you learn, it's going to move. It's never in the same place twice, and it's different for everyone."

My very first magazine assignment, back in 2005, was to cover the RoboCup U.S. Open in Atlanta for *Popular Mechanics*. RoboCup was established in 1997, the same year that IBM's Deep Blue computer beat Garry Kasparov, and its stated goal is to "develop a team of fully autonomous humanoid robots that can win against the human world soccer championship team by the year 2050." The marquee event in 2005 was a four-aside clash between teams of small, vaguely Snoopy-like robot dogs called AIBOs, sold as toys by Sony.

What amazed me most about the AIBO soccerbots was how spectacularly bad they were. In many games, the competitors just stood around scanning their heads back and forth. When they decided to move, they sometimes wandered off the field entirely, earning thirty-second penalties. I'd naively assumed that, even if the subtleties of the beautiful game eluded them, fundamentals like kicking the ball would be fairly straightforward to program. But the dogs were struggling even with basic tasks like walking, and the move from various home labs to the competition venue at Georgia Tech, with different lighting and unfamiliar objects in the background, had thrown off their ability to spot the orange ball. Even worse, the free T-shirts given out by the tournament hosts were orange like the ball. At one point, the referee had to ask a kid wearing the tournament T-shirt to hide behind a barrier because dogs kept charging off the field toward him.

At around the same time, across the Atlantic at the Sony Computer Science Lab in Paris, a team of researchers was trying a very different approach to training AIBOs. An engineer named Frédéric Kaplan had brought the very first AIBO prototype to Europe. "That was actually a whole adventure to pass through customs with that thing," he later recalled. Sony wanted the AIBO to function like a real pet, learning and forming a personality based on the unique life trajectory it shared with its owner. But—as the RoboCup researchers had also discovered—teaching robots how to make sense of the world just by analyzing the flow of incoming sensory information is way harder than it looks. Kaplan spent days on end showing the AIBO simple objects like a ball or a plastic cucumber, under different lighting conditions and on different backgrounds. "You would say 'Ball, ball, ball,'" Kaplan recalled grimly. "And then you would ask, 'What is it?' And it would say, 'Oh, cucumber. . . .' It was very exhausting."

So Kaplan and a colleague named Pierre-Yves Oudeyer decided to try a radically different approach—one that sought to endow the AIBOs with the childlike wonder of, well, a child. Human development is, after all, "the only known process that reliably produces human-level intelligence," as one paper on the development of exploratory behavior puts it. Instead of teaching the AIBO, they decided to let it discover its own capabilities. The only guidance they gave it was that it should make predictions about the effects of any actions it took, monitor the effects to see how good its predictions were, and preferentially choose actions that produced the biggest reductions in prediction error. In other words, they programmed it to follow exactly the same slope-chasing tactic that generates exploratory behavior in the predictive processing framework.

Videos of Oudeyer and Kaplan's experiments show an AIBO on a colorful play mat, exploring the dangling toys of a baby gym. The AIBO makes some repetitive but unintelligible noises, like the babbling of a baby, as it explores its vocal capabilities. And it swivels its head, opens and shuts its mouth, and moves its limbs back and forth—what development biologists and roboticists call "motor babbling"—as it learns about its own physical capacities. Occasionally, its limb smacks one of the toys, making it swing back and forth. That's a big prediction error—probably too big in the early stages of learning, when it's still figuring out what its limbs do, but something it will later find fascinating.

Oudeyer and Kaplan ran and reran the experiments thousands of times, looking for patterns in the learning trajectories. Each time, the AIBO dog would start with random movements. Then, by chance, it might figure out how to, say, walk backward. "And then our robot would go for one hour backward," Kaplan said. Eventually, walking backward was no longer surprising, "and then he would discover that a small shift in his parameter would go a bit on the left." So it would walk to the left for an hour. And so on. Oudeyer and Kaplan published the details of their intrinsically curious AIBOs in 2005. The initial generation of AIBOs didn't live up to Sony's hopes, and production was canceled the following year. RoboCup has moved on and now uses other robots for its competitions. But when Sony announced the relaunch of the AIBO line in 2017, it included a new feature, based on Oudeyer and Kaplan's research: artificial curiosity.

Whether this model of artificial curiosity—based on reducing prediction error—is really what drives humans remains to be seen. The success of Oudeyer and Kaplan's AIBOs, along with many other similar robot and AI systems in the years since, provides one proof-of-concept. There's also growing evidence that the human reward system is more nuanced than the original picture of dopamine as a simple signal of reward or prediction error, Miller says. Instead, dopamine and other neuromodulators may play a broader role encoding the precision of your predictions, consistent with the idea that they enable your brain to track whether you're reducing prediction errors more quickly or more slowly than expected. And then there's the behavioral evidence: How else can I explain how I ended up in the woods—and among the thistles—armed only with a compass, a map, and my running shoes?

The reason the military likes training its recruits in orienteering isn't just so they don't get lost on the battlefield, though that's certainly a useful byproduct. The real benefit, West Point's Hendricks told me, is that "you learn the effects of stress and fatigue on your decision-making." The West Point team uses GPS and heart-rate data to analyze their performance after training and competitions. "You can determine that, yeah, at that point when your heart rate was really high for a certain period of time—bang, that's when your mistake was made," he said.

That certainly proved true for Adrian, Nevin, and me at my first Raid the Hammer. As hours passed and the race wore on, our decisions got worse and worse. Between one pair of checkpoints, we passed the same team—an older, slower trio—three different times, sprinting past them only to make yet another boneheaded navigation error a few minutes later. Late in the race, we found ourselves locked in a tight battle with a team of thirteen-year-old girls. They were participating in the shorter junior version of the race, which had started an hour later but finished along the same route. Even though we were probably running 50 percent faster than they were, we kept meeting at successive checkpoints then heading back into the forest in opposite directions—they in the right direction, we in the wrong. Only the shredded remnants of our pride prevented us from giving up and simply following them from checkpoint to checkpoint.

That memory floated to the surface more than a decade later, in 2024, when I read a newly published study in *PLOS One* about the cognitive benefits of orienteering. Scientists at McMaster University had found that even a single session of orienteering was enough to boost levels of brain-derived neurotrophic factor (BDNF), a key neurochemical that promotes brain plasticity and improves performance in a spatial memory test. The combination of physical and mental stimulation seemed to be more effective than the same intensity of purely physical exercise. The lead author was a graduate student named Emma Waddington, a veteran of Canada's national orienteering team—and, as it happens, one of the thirteen-year-old girls who'd schooled me and my teammates in 2011.

The necessity of engaging my brain and body together was one of orienteering's big selling points for me, and not just because I wanted to boost my BDNF levels. By that point in my life, running had become more of a habit than an adventure—and habits are, as the Israeli neuroscientist Moshe Bar puts it, a double-edged sword. Forming habits allows us to automatize frequently repeated actions to save time and energy; but it's also "the same mechanism that prevents us from enjoying an éclair as richly every time we have one," Bar says. In the predictive processing framework, we construct reality from the intersection between the top-down predictions generated by our brains and the bottom-up sensory information collected by our senses. Habits skew toward the top-down predictive side; I wanted to start noticing the world around me again.

Bar, with his colleague Noa Herz, has proposed that we have two fundamental and opposing states of mind, which he calls "exploratory" and "exploitatory." In an exploratory state of mind, you're giving more weight to incoming sensory data from the real world than to internal predictions based on past experience. You tend to have a broader field of attention, more flexible thinking, more positive mood, and (as the name suggests) are more likely to explore. The purest form of this exploratory state of mind, Bar says, is mindfulness, in which you observe the world around you unimpeded by internal judgments and expectations. An exploitatory state of mind, in contrast, relies more on the brain's preconceived predictions. Your attention is narrower and more focused, but you're less creative, less tolerant of uncertainty, and in a more negative mood. On this end of the spectrum, the extremes are dreaming and daydreaming, in which sensory input is absent or minimal.

Most of our time, of course, is spent somewhere between the two extremes, as we balance predictions and observations. When I'm stressed-out about an impending deadline or rehashing a recent interview in my head rather than listening to what my wife and kids are saying at the dinner table, I'm skewing top-down. When I'm maneuvering my canoe through a tricky set of rapids, or otherwise immersed in some engrossing activity, I'm more bottom-up and attuned to my surroundings. The psychologist Paul Bloom, in his book on the pleasures of suffering, quotes a dominatrix on this point: "A whip is a great way to get someone to be here now. They can't look away from it, and they can't think of anything else."

Both the exploratory and exploitatory states of mind are useful in the right context. If you're feeling happy and curious, it might be better to seize the moment to do some brainstorming for a new project, and save your expense reports for another time when you're less enthusiastic but more focused. Bar and his colleagues have also found that it's surprisingly easy to shift people's states of mind. One set of experiments involved reading chains of word associations that were either narrow (dog-cat-puppy-animal-friend-house-food-biscuit . . .) or broad (dog-bone-chicken-rooster-farm-cow-milk-cookies . . .). Simply reading the broad lists put people in a more exploratory frame of mind and improved their mood. More generally, according to Bar, putting yourself in a new environment, trying a new dish, or—yes!—going for a run can loosen the shackles of your brain's predictive machinery. "Sometimes," he writes, "just ice cream and a funny movie can do the trick."

All of these hacks assume that you *want* to explore more. In the fall of 2011, Raid the Hammer was at the peak of my Wundt Curve, drawing on my existing skills but also forcing me to learn new ones and see my surroundings differently. In the years that followed, I did more orienteering races, tried other types of adventure races, and dabbled in mountain running. This exploration was a ton of fun and rekindled my love of running. But after my kids were born in 2014 and 2016, I found that I was getting my fill of prediction error as a parent. I retreated to familiar 5K road races for a few years, and found that just as satisfying. Then, as my kids got older, I started to feel the itch again. I ran a few seasons of masters cross-country races with a team of friends and began to sense the draw of Raid the Hammer again. The sweet spot is a moving target.

Any organism that follows the slope-chasing imperative of the free energy principle will be inherently restless. "Every time it figures out a niche, it loses the prediction error that's requisite to it continuing to improve," Miller points out. "So it goes seeking again for other good errors to digest." But what constitutes a digestible prediction error remains highly personal. José Ruvalcaba's Mexican hometown, Ameca, has a population of sixty thousand, many of whom are presumably glad they *didn't* choose to migrate to the United States. Having an inborn urge to explore, as I claim we all do, doesn't mean we always have to be exploring. In the coming chapters, we'll delve into the mystery of how we decide when to explore, what strategies we use when we do, and what we get right and wrong in the process.

Part II

HOW WE EXPLORE

4

EXPLORE VS. EXPLOIT

On May 3, 2019, Nils van der Poel completed his military service. The twenty-three-year-old had just finished a year of special forces training with Sweden's elite Army Ranger Battalion, based in the remote subarctic town of Arvidsjaur. He took a day off, and then went for a run, covering fifty kilometers—just over thirty-one miles—in four hours and seven minutes. A few days later, he went for a five-hour bike ride, then followed that with a six-hour ride the next day, then a pair of three-and-a-half-hour runs. One of the most radically unconventional experiments in the annals of Olympic sport—one whose fruits wouldn't become apparent until almost three years later—had just begun.

Van der Poel grew up in a midsize Swedish town called Trollhätten (the name means "troll's hat," after a local waterfall), playing a Northern European version of ice hockey called bandy. When he was eight, the coach of a local speedskating club that shared the same rink noticed that he had a Dutch last name, from his paternal grandfather, and convinced him that this meant he was destined to speed skate. The coach was right, it turned out, and Van der Poel won gold medals at the World Junior Championships in 2014 and 2015. After the first victory, he was elated; after the second, relieved. The letdown bothered him, and he took a six-month hiatus from the sport. He came back to qualify for the 2018 Olympics in Pyeongchang, but his fourteenth-place finish there left him even more underwhelmed. At the end of that season, he walked away from the sport and into the army.

The ascetic life of an elite athlete had given Van der Poel plenty of time to think, and the conclusion he came to was that he didn't like that life. He had given up so much in order to pursue his athletic dreams—and the worst part was the nagging suspicion that some of the sacrifices were unnecessary. In addition to time on the ice, speed skaters spend endless hours doing other forms of training: one-legged squats, Rollerblading, slide board drills, running sideways on treadmills, not to mention stretching and lifting weights. Van der Poel wasn't convinced any of it mattered. Elite athletes also tend to be high-functioning exercise addicts, loath to miss even a single workout. Van der Poel wanted a few days completely off ... *every week*.

Not all speed skaters do exactly the same training, of course. But there is a broad set of principles that they virtually all adhere to, in large part because that's how the current champions and their predecessors got to the top. If you want to try something new, that inescapably means *not* doing the things that have already proven to work. This, in a nutshell, is what scientists who study decision-making refer to as the "explore-exploit" dilemma. You might intuitively assume that the opposite of exploring is, say, lazing around on the sofa with Netflix and a family-size bag of Cheetos. But that's really the difference between action and inaction—between sticking with speed skating and quitting entirely, for Van der Poel. For decision scientists, the more interesting distinction is whether you try something new or stick with what's familiar—that is, *explore* a new path or *exploit* your existing knowledge.

This terminology was first proposed by a British mathematician named John Gittins in the 1970s, and later popularized by American management guru James March. And once you're aware of it, you start noticing explore-exploit decisions everywhere in your life. Should you watch the next season of a show that you thought was okay, or scroll through your Netflix recommendations to find a new series that you might enjoy more—though you might fritter away a Friday evening before realizing that you hate it? Should you spend another spring break in Florida, or try Costa Rica? Should you keep plugging away at your career as a banker in Ameca, or cross the border to try to get rich? The dilemma isn't restricted to humans: a foraging bee weighs how long to stick with its current patch of flowers before looking elsewhere; a multinational corporation shifts resources back and forth between marketing its current product and researching possibilities for its next one.

Van der Poel didn't want to keep doing what he'd done in the past, and what all his rivals were doing; he wanted to try something genuinely new. In 2018, he began talking with a coach named Johan Röjler, himself a former world junior champion and three-time Olympian, and they came up with a four-year plan to prepare Van der Poel for the 2022 Olympics. It began with the year of army training as a sort of mental and physical reset. Then starting in May 2019, he spent a year running, biking, cross-country skiing, and ski mountaineering. It wasn't a vacation, though. A few months into the year, he ran for twenty-three hours over two days along the King's Trail, a mountainous route in Swedish Lapland. He had to cut the next day's run short after a mere four hours because his ankles were so sore. A few weeks later, he racked up more than fifty hours of running in five days in a trail stage race. He spent ten and a half hours climbing stairs for a total ascent matching the height of Mount Everest, and eventually logged single days with over a hundred miles of running and more than five hundred miles of cycling, each taking more than twenty hours. During this year, he settled into a pattern of training insanely hard from Monday to Friday then taking the weekend completely off, which accommodated his new-found passion of skydiving. The one thing he almost never did: put on his skates.

After a year of building aerobic fitness, having racked up twenty ultramarathons and biked the length of Sweden (not to mention completing a thousand skydives) Van der Poel was finally ready to think about speed skating again. Here too he and Röjler had new ideas. His best events were the 5,000 and 10,000 meters, lasting roughly six and thirteen minutes respectively. To win the Olympic 5,000, he would need to skate laps of the 400-meter oval in roughly twenty-nine seconds; for the 10,000, he would need thirty-second laps. He decided to train with the goal of accumulating as many laps as possible at exactly those paces, mimicking the demands he would face in races. To do so, he divided the training year into three components. The aerobic season involved the running and biking and other activities he'd spent the previous year doing, with a goal of as much as thirty-five relatively easy hours per week. The threshold season involved biking at an intermediate intensity for about twenty-five hours per week. Then in the specific season, which began only three weeks before the first World Cup race of the season, he finally moved onto the ice, where he would use his painstakingly accumulated fitness to execute endless laps exclusively at race pace.

Training like this was a huge gamble, as was his entire four-year odyssey since the previous Olympics, and the speed skating world didn't know what to make of it. "I have trained many athletes," Sweden's head of Olympic talent development, Wolfgang Pichler, said. "Man, I've never seen anyone like him before." In the lead-up to the 2022 Olympics, Sweden's national public broadcaster, SVT, ran a documentary titled *Nils van der Poel: Genius or Fool?* Van der Poel himself, on the other hand, had never felt more confident. "Some said my plan was foolish, and my decision raised some eyebrows," he told SVT, "but to me it felt logical and smart."

The crux of exploring, in the explore-exploit framework, is that by definition you're *not* choosing what superficially appears to be the best available option. If you take all the knowledge you currently have, sift it, weigh it, and select the objectively best alternative, that's simply exploiting. In contrast, exploring involves picking a path that looks less promising but is also shrouded in uncertainty. That uncertainty is what you're seeking, because in resolving it you'll gain knowledge about the world and possibly discover a better path. The knowledge isn't free, though. If you love the butter chicken at your favorite Indian restaurant but decide to give the biryani a shot for a change, the most likely outcome is that you'll be disappointed. And if you're a former world junior champion entering your athletic prime, adopting a radically unorthodox training program means giving up a decent shot at picking up an Olympic medal with the conventional approach. You're passing up good for the possibility of great.

Ordering dinner and training for the Olympics are very different endeavors with very different stakes. But they both map onto the framework of a classic gambling problem called the "multi-armed bandit." The terminology was introduced in the 1950s by Harvard researchers Robert Bush and Frederick Mosteller, who were studying decision-making: subjects in their experiment had to choose between two buttons on a machine they dubbed the "two-armed bandit," in analogy to the one-armed bandits—that is, slot machines—in casinos. In the scientific literature on explore-exploit decisions, so-called bandit problems have become the common currency for testing how we make decisions under uncertainty, how we *should* make those decisions, and what factors tip the balance in one direction or the other. You can express any real-world explore-exploit scenario as a bandit problem; and the mathematical insights we gain from studying bandit problems can—at least in theory—be applied back to those real-world scenarios.

Here's a simplified example: imagine walking into a casino that contains two slot machines. Each of these machines, you're told, has been programmed to pay off with a certain probability. You might win 20 percent of the time, or 50 percent of the time, or 80 percent of the time—but you don't know the settings of each machine in advance. You're given a hundred tokens to play the machines. How do you proceed?

The best strategy is fairly intuitive, at least in broad strokes. You should spend some tokens on the first bandit to get a sense of how often it pays off. And you should spend some tokens on the second bandit to get a sense of how often *it* pays off. Then, once you know which one is better, you should spend the rest of your tokens there. That's the easy part. The hard part is figuring out how many tokens you should devote to exploring each bandit before you switch to exploiting your knowledge. If one pays off 90 percent of the time and other just 10 percent, you'll know pretty quickly. But if it's, say, 40 percent and 30 percent, it will take a lot longer to be confident about which one is better. Figuring out exactly how long to explore turns out to be fiendishly difficult. "The problem is a classic one," Cambridge University statistician Peter

Whittle quipped in 1979. "It was formulated during the war, and efforts to solve it so sapped the energies and minds of Allied analysts that the suggestion was made that the problem be dropped over Germany, as the ultimate instrument of intellectual sabotage."

Whittle's comments were prompted by the news that a solution had been found. In the mid-1970s, Unilever, the British consumer goods conglomerate, asked John Gittins to help it optimize how it screened various chemical compounds as potential drugs—a riddle, it turned out, that was a multi-armed bandit problem in another guise. Gittins came up with a mathematical technique that assigns a value—what's now known as the Gittins index—to each "arm" in a bandit problem at any given point in time. Always choose the arm with the highest Gittins index and you'll maximize your eventual payoff.

The classic real-world incarnation of a bandit problem—one where we've all encountered both the joys and the perils of an exploratory choice—is ordering food. Let's say, for example, that you're comparing restaurant options. You've tried Alice's Pizza ten times, and enjoyed it six of those times; you've been to Bob's Burgers twice, and got one good meal. Your first estimate, then, is that you have a 60 percent chance of satisfaction at Alice's and 50 percent at Bob's. Pizza it is, if you're following the "greedy" approach of always selecting the option with the highest (according to your current knowledge) reward—if you're exploiting, in other words.

But you haven't really given Bob's place a fair chance, since you've only been there twice. How do you know if it's worth devoting more time to exploring there? The Gittins index offers a way of comparing your options that takes into account how complete your information is. So you grab your copy of Gittins's 1989 textbook, *Multi-Armed Bandit Allocation Indices*, and flip to the appendices at the back where exhaustive tables of Gittins indices are compiled for various scenarios. With a few assumptions, you find that with six wins out of ten attempts, Alice's has a Gittins index of 64.0 percent, while Bob's, with one win in two attempts, is at 63.5 percent. In other words, it's pretty much a toss-up between the two restaurants, thanks to how little we know about Bob's cooking skills. If you stick with Alice's until you've had twelve good meals in twenty visits, her raw batting average remains the same but her Gittins index slips to 62.4 percent. At that point, it's time to switch from exploiting to exploring and head over to Bob's.

The Gittins index, in various forms, has been deployed to tackle very specific and narrowly defined problems in economics, engineering, military planning, website design, and numerous other fields. But it's not a general theory of how to explore. Modern computer programmers don't program their AI bots to make choices with a Gittins index, because the math is too unwieldy and the assumptions too unrealistic. And neuroscientists don't believe that we're running Gittins calculations in our brains when we order dinner. Instead, scientists have come up with a set of "heuristics": rough-and-ready exploring strategies that aren't perfect but get you pretty close to the best possible outcome at a fraction of the computing cost. These heuristics have become crucial to the success of machine-learning systems in computer science—and surprisingly, they seem to mirror how our brains make explore-exploit decisions too.

Let's go back to the scenario where you're choosing between two slot machines, knowing nothing about either of them. You try each one once; one of them wins and the other loses. You now have a crude estimate that one machine will win 100 percent of the time and the other will never win. The greedy strategy suggests you should play the winner . . . and then play it again, and again, and again. In fact, you'll never return to the other machine. Even if you lose the next ninety-nine pulls in a row, you'll still estimate that the first machine has a one percent probability of winning, which is better than zero percent. The greedy strategy is making you blind to the possibility that the other machine might be pretty good despite having dealt you an initial loss. In real life, of course, your intuition will tell you that you shouldn't completely dismiss a bandit just because it lost once. That intuition is, in effect, your exploring algorithms in action.

There are two broad categories of exploring strategy that prompt you to deviate from the greedy strategy: random exploration, and uncertainty-directed exploration. Random exploration is exactly what it sounds like. Sometimes, instead of selecting the bandit with the highest reward, you should simply flip a coin. That will help you avoid getting stuck in a rut where you think you've found the best option: you'll occasionally try other options purely at random, and sometimes these deviations will reveal that another bandit is better than the one you're currently exploiting. You can tweak this random algorithm to make it a bit smarter. For example, you could start out devoting 30 percent of your turns to random exploration. Then, as you gain knowledge, you could gradually reduce the amount of random exploration to 20 percent, then 10 percent, and eventually even to zero once you have a good sense of how rewarding each bandit is.

The other main exploring strategy is uncertainty-directed exploration, which involves preferentially sampling the options you know the least about. This is the intuition, in the example above, that tells you that a bandit that you've only tried once is worth trying again, even if you lost on the first pull. It's also the type of exploring that emerges naturally from the free energy equations in the last chapter: if your goal is to minimize surprise in the long term, you'll want to check out the options you know the least about to roust out any big surprises that might be lurking there. In practice, computer scientists have a simple way of programming uncertainty-guided exploration into their bandit-playing systems: they add an "uncertainty bonus" to the potential payout of each option to make less-traveled paths more attractive. That's effectively what the Gittins index does: Bob's index of 63.5 percent represents a baseline success rate of 50 percent (one good meal in two tries) plus a 13.5 percent uncertainty bonus. The size of the uncertainty bonus depends on how much you know about a given option: the less you know, the bigger its bonus.

All of this sounds very abstract. Amazingly, though, when you ask people to play multi-armed bandit games, their choices clearly reveal the presence of both random and uncertainty-directed exploration. You can tweak the parameters of the game to ramp up one or the other. When some choices are much more uncertain than others, we use more uncertainty-directed exploration, which is mediated by dopamine's prediction error tracking. When there's a lot of uncertainty about all the choices, we tend to exhibit more random exploration, which is implemented in the brain in a very literal way: random variability in the signals traveling from neuron to neuron ensures that you don't *always* make the same choice when faced with a given set of circumstances.

Animal experiments suggest that we can tune this level of "noise" in our brain signals, depending on the context. When zebra finches are learning to sing, for example, they engage in characteristic babbling that is often compared to the babbling of human babies. "There's a piece of the brain that generates randomness for the purpose of learning," Massachusetts Institute of Technology neuroscientist Michale Fee explains. When they're adults, male zebra finches continue to harness randomness to explore their vocal capabilities and fine-tune their songs. When no females are around, the notes and timings of males' songs are more variable, as are the neural signals that produce the songs. But if there's a female in the audience, a burst of the neurotransmitter norepinephrine shuts down the noise in the neural circuits, and the males deliver a highly polished and repeatable performance. The difference between practice and performance, in other words, is that the former incorporates some hardwired random exploration while the latter is pure exploitation of what you've learned so far.

Nils van der Poel reemerged from the speed skating wilderness in December 2020, when he entered a small competition—his first since 2018—in the German mountain town of Inzell, where he was training at the time. He won the 10,000 meters, but that's not saying much: he was the only entrant. Still, the performance qualified him for the following month's European championships, where he shocked the favored Dutch by picking up a silver medal in the 5,000 and winning the 10,000. "After a two-year hiatus, Nils van der Poel, twenty-four, has made a sensational comeback—and is suddenly one of the world's best skaters," a Swedish newspaper marveled. And the best was still yet to come. At the world championships a few weeks later, Van der Poel won both the 5,000 and 10,000, setting a world record in the latter event and winning by a staggering margin of almost thirteen seconds.

Commentators were baffled. When they last saw him, Van der Poel had been a good skater who'd finished fourteenth at the Olympics. He then disappeared for two years and had now returned as the dominant figure in long-distance skating. With a hint of mischief, Van der Poel cited Therese Johaug, a Norwegian cross-country skier who dominated her sport after missing eighteen months with a doping ban, as his inspiration: "I got the idea that if you want to achieve the best possible results in the long term, it is probably not beneficial to have peaks of form every year. It might be better to have a really long basic training period."

Whatever the explanation, Van der Poel was suddenly a favorite for the 2022 Olympics—and the preparations began immediately. His world-record 10,000 race was on February 14, 2021. When it was done, he put away his skates—and then didn't put them on again until October 4, almost eight months later. Instead, he spent the spring and summer once again running, biking, ski mountaineering, and generally racking up dozens of hours of aerobic exercise each week. In August, he switched to threshold rides on his bike. And in October, he began once again skating endless laps at race pace. In December, he set a 5,000-meter world record at a World Cup race in Salt Lake City. By the time he arrived in Beijing for the Olympics, he was no longer flying under the speed skating world's radar.

Of all the strange and unexpected parts of Van der Poel's journey, the most surprising—and the one that turned him into a folk legend among endurance athletes around the world—is what he did after his final race in Beijing. He had begun his Olympic campaign with a nail-biting last-lap come-from-behind victory over Dutch skater Patrick Roest for the 5,000-meter gold on February 6. Then he uncorked another world record to dominate the 10,000 meters, winning once again by thirteen seconds, on February 11. The very next day, a website called howtoskate.se went live, containing exactly one document: a sixty-page pdf containing a personal, funny, impassioned, reflective, and supremely unorthodox training manifesto, along with a complete record of his training logs from May 2019, when he was discharged from the army, right up to the Olympics. It began with a quote from Carl Jung: "It seems that all true things must change and only that which changes remains true."

Most Olympic champions keep their training a closely guarded secret; or if they share details on public platforms like Strava, it's fragmentary and incomplete, stripped of its context and underlying rationale. Van der Poel, in contrast, seemed eager to reveal it all, and more importantly, to explain his thinking. And if it helped his competitors? All the better. "I also wish for the sport to keep developing and for my records to be broken," he explained. "For those who might want to, I wrote this document."

But there was more to it than a selfless desire to advance the sport. Van der Poel had taken a huge risk, wandering far off the established route to success and putting himself through enormous hardships. "I kept in mind that when I set out to break a world record, I set out to achieve what no one else had ever achieved before," he wrote. "I understood that I would not fulfill that goal using means others had already undertaken." Explorers' journals are a long-established literary genre, a line that runs from ancient Greek voyagers like Pytheas through Christopher Columbus and Charles Darwin all the way to the Instagram feeds of modern adventurers. Now Van der Poel wanted to join that tradition, mapping out the details of his voyage, filling in the blank spots on the map, and pointing out where dragons might still lurk. In an interview with Stephen Seiler, a prominent American-born sports scientist based in Norway, he tried to articulate his sense of discovery: "I felt like I had seen a way to do the sport that no one else had noticed. And I felt like it was standing in front of me. Like, why hasn't anyone else done it this way? This seems to be a lot better."

The multi-armed bandit paradigm strips down explore-exploit decisions to their bare components. That's useful but doesn't really capture the complexity of real-world decisions like the ones facing Van der Poel. In recent years, scientists have begun looking outside the lab for datasets that reveal how we handle these sorts of decisions in the wild.

Even the classic example of ordering food is much more nuanced in reality than the simple choice between Alice's and Bob's restaurants that we considered above. In 2018, a research team led by Eric Schulz, who now heads the Computational Principles of Intelligence Lab at the Max Planck Institute for Biological Cybernetics in Germany, partnered with the food delivery company Deliveroo to obtain full data from 195,000 customers who placed 1.6 million orders to thirty thousand restaurants in 197 cities during February and March of that year. Amazingly, the patterns in the dataset show telltale signatures of some of the exploring algorithms theorized by mathematicians and previously only observed in carefully controlled bandit studies in the lab.

You're probably looking for several different things when you scroll through a food delivery app. You want the food to be tasty, and perhaps healthful. You want it to arrive at the right time, and to be priced fairly. The app gives you hints about all these factors: food styles, prices, delivery times, user ratings. Some of the patterns in the data are obvious: all else being equal, diners prefer to try less-expensive restaurants with faster delivery times and higher ratings. None of those clues are totally reliable, though, so you're forced to decide among a wide range of options with imperfect knowledge about the likely outcomes—the perfect ingredients for a challenging explore-exploit dilemma.

The bad news for culinary explorers is that, as predicted, exploration has a cost. The average rating when diners tried a restaurant for the first time was slightly lower than when they returned to a previously sampled restaurant, at 4.26 versus 4.52. To put it another way, old favorites produced five-star meals 70 percent of the time while new choices managed it less than 60 percent of the time. The good news, though, is that exploration pays off in the long run: as people accumulated more orders, their average ratings crept up as they discovered better options and discarded the duds.

As for random versus uncertainty-directed exploration, the former isn't a great way to order food. I once tried this approach as a young backpacker in Spain, ordering lunch in a tiny seaside village where no one spoke English. I knew just enough Spanish to recognize the *mariscos* section of the menu—it seemed like a good bet, given the fishing boats moored nearby—so I pointed at something and hoped for the best. What eventually arrived, more than an hour later, was a whole boiled octopus, head, tentacles, and all, completely unseasoned, accompanied by not so much as a sprig of parsley. I make it a point of pride to always clean my plate, and I sawed away at the tasteless, rubbery beast for as long as I could but eventually had to surrender due to jaw fatigue.

Uncertainty-directed exploration, on the other hand, makes sense, and there are several smoking guns in the Deliveroo data to indicate that diners were implicitly including an uncertainty bonus in their deliberations. If you look at the number of ratings each restaurant has received, you find that—all else being equal—more ratings make it *less* likely that a diner will choose to explore that restaurant for the first time. This is counterintuitive. If you're using the ratings to maximize your immediate chances of a good meal, more ratings should give you greater confidence that you won't get a subpar meal. The flip side, though, is that you're also less likely to get an unexpectedly spectacular meal. If thousands of people have already rated the place, it's probably not an undiscovered gem. A restaurant with few ratings, then, signals a rich source of uncertainty, and this attracts people to explore it.

Another source of uncertainty is the variance in restaurant ratings. Consider a restaurant that almost everyone rates as three stars, with just a handful of twos and fours. Compare it to another restaurant that has an identical three-star average, but with wide variance: lots of twos and fours, and even some ones and fives. It turns out that, even after sampling each restaurant once, customers were more likely to reorder from the restaurant with wider variance—as you'd expect if we're drawn to the possibility of gaining information to resolve uncertainty.

So how do all these pieces fit together? Schulz tries out various restaurant-choosing algorithms to see which one best matches the actual behavior of the human diners. A model that simply tries to optimize price, average rating, delivery time, and number of past ratings does pretty well. But one that does all that *and* adds an uncertainty bonus doubles its predictive power. The particular form of that uncertainty bonus turns out to be significant: it's called the Upper Confidence Bound algorithm, and it's sometimes referred to, in the multi-armed bandit literature, as "optimism in the face of uncertainty."

The main goal of the Upper Confidence Bound algorithm is to minimize regret, which in this context is a mathematical quantity defined as the difference between the rewards you got and the rewards you could have had if you'd magically made all the right decisions. It's what might have been. In reality, no one is clairvoyant, so regret always increases with the passage of time (an observation that transcends math). But in 1985, Columbia University mathematicians Tze Leung Lai and Herbert Robbins showed that you can keep regret to minimum by following a relatively simple rule that boils down to always choosing the option with the biggest potential upside. The confidence bound is basically the error bars on your estimate: if you think a slot machine pays off 60 percent of the time, plus or minus 10 percent, its upper confidence bound is 70 percent. A less-explored option that pays out 50 percent of the time plus or minus 25 percent would have an upper confidence bound of 75 percent, and be the preferred option in Lai and Robbins's algorithm. In practice, of course, we rarely make decisions by explicitly calculating expected outcomes, much less the associated error bars. But the general principle is straightforward: act optimistically, pursuing the options with the greatest (realistic) upside, and you'll leave yourself with fewer regrets.

One of the most important differences between the simplest versions of the multi-armed bandit problem and the real world is that rewards in the real world keep changing. You may have tried all the burger joints in town and settled on a favorite, but it would be a mistake to assume that you should never again try a burger anywhere else. Maybe one of the other joints has hired a brilliant new burger chef. To simulate this behavior, scientists run "restless bandit" experiments, in which the rewards offered by each slot machine can shift up or down over time. This scenario increases the value of continuing to explore.

Longline fishing boats in the Gulf of Mexico got a stark demonstration of this principle when regulators imposed an emergency six-month shutdown of their prime fishing grounds in 2009 because too many loggerhead turtles were being accidentally killed. Researchers at the University of California, Davis crunched the data from all 106 longline boats in the region for a two-year period leading up to the shutdown. Under normal conditions, there was no clear relationship between revenue and measures of exploratory behavior. In a sufficiently complex environment, there's not necessarily a single "right" answer to the explore-exploit dilemma. Those who explored more may have found richer fishing grounds some of the time, but at the expense of fuel costs and time spent fishing at grounds that turned out to be duds. With decades of experience, each boat had found its own balance, and they all produced similar results.

But when the shutdown hit, the situation changed. Almost half the boats simply stopped fishing, and those that stopped tended to be ones that had less exploratory fishing patterns. Presumably they had less knowledge of alternate fishing grounds outside the prime area that had been closed. Among the boats that kept fishing, revenue was highest among the most exploratory boats. The information that the explorers had been accruing over the years didn't give them any obvious advantage under business-as-usual conditions, but it proved to be highly valuable when the status quo was disrupted—an intriguing finding, the researchers point out, in light of the increasing frequency of climate-related upheaval around the world.

Another natural experiment took place on the evening of February 4, 2014, when London Tube workers began a forty-eight-hour strike. Participation in the strike was voluntary, so disruptions across the network depended on how many workers showed up in each place: a total of 171 out of 270 Tube stations were closed for at least a portion of the following two days. As a result, commuters scrambled to find alternate routes—and, in some cases, discovered that their new routes were superior to the ones they'd been using all along. "The walk from Liverpool Street was a refreshing change from the horrors of the Circle Line," one commuter told the BBC. "I suspect I may permanently switch so I can cut out the most stressful part of my journey."

This wasn't just an anecdotal finding. Researchers from Cambridge and Oxford, along with the International Monetary Fund, sifted through anonymized transit data to identify more than seventeen thousand people whose morning commute was regular and predictable: they used the Tube every morning in the twelve weekdays leading up to the strike, at least once during the strike, and every morning in the six days following the strike. During the strike, almost half of them started their journey at a different Tube station than usual, and more than half of them finished at a different station. This might not be as bad as it sounds. The Tube network, with its eleven lines, is notoriously complex: there are at least thirteen reasonable ways to travel between King's Cross and Waterloo stations, the researchers point out. The standard system map severely distorts the actual distances between stations. Tourists often don't realize that Covent Garden and Leicester Square are less than three hundred yards apart, for example, so they take the twenty-second Tube trip. And time and distance aren't the only factors: some routes are more crowded and unpleasant than others.

Sure enough, among the most rigidly habit-bound commuters—those who entered and exited at the same stations *every* morning prior to the strike—about 5 percent of them switched either their entry or exit point in the days following the strike, presumably because they'd found a better route. This is a remarkable result, because it suggests that endlessly repeating a task isn't enough to ensure that we find the best way of doing it. By one estimate, about 45 percent of the things we do are repeated daily; we don't wake up every morning trying to optimize those habits from scratch. In the 1950s, Herbert Simon coined the term "satisficing": faced with the irreducible complexity of the real world, it's often best to settle on a solution that's good enough rather than exhaust yourself pursuing the marginal benefits of the best solution.

Sometimes, though, we settle too soon, or get so locked into habits that we fail to rethink them when conditions change. The researchers compare their Tube results to the Porter Hypothesis, a controversial idea advanced by Harvard Business School professor Michael Porter in 1991. Porter argued that strict environmental regulations, far from stifling innovation, would actually force industries to explore better, more efficient, and ultimately more profitable ways of running their businesses. Few companies welcome stricter regulations, and no commuter looks forward to a transit strike. But if the world forces you to shake up your routine, make the best of it by looking for new ways of attacking old problems.

As versatile and ubiquitous as it is, not all exploring decisions map onto the multi-armed bandit model. There's another class of explore-exploit dilemmas that mathematicians call optimal stopping problems, illustrated most famously by a scenario that Merrill Flood presented at a conference in Washington, DC, in 1950. Flood was a giant in the field of operations research, which uses math to help people and organizations make better and more efficient decisions. These days he's best remembered for his early work on the Traveling Salesman Problem, which attempts to find the optimal route through a list of cities. He also formulated the Prisoner's Dilemma, the most famous problem in game theory, and—by some accounts, at least—he coined the word software.

But the problem he posed in 1950 was more personal. Suppose you're in love and trying to figure out whether your current partner is, in fact, the one. He's the greatest love of your life so far. But how do you know if, by exploring for a little longer, you might meet someone better? Flood called this timeless quandary the Fiancé Problem, and he didn't have a mathematical answer for it—yet. But he had a reason for posing it. His daughter, Sue, had just graduated from high school and was dating a much older divorced man with two children. The relationship seemed to be getting serious, to the chagrin of Flood and his wife. Sue was in charge of taking notes for the official minutes of the conference, and Flood hoped that seeing the pursuit of love presented as a logic problem to be optimized might convince her that she needed to spend a little more time exploring.

Optimal stopping problems can take on many different forms. On your way to a concert, how many open parking spots should you pass before you take one and walk the rest of the way? When you're selling your house, how do you decide when to accept an offer rather than holding out for a better one? If you're successfully robbing banks, how many more should you hit before donning a rubber nose and fake mustache and settling into retirement? The right answers depend on the details of the problem and the constraints you impose. In the simplest version of the Fiancé Problem—or, as it became better known in the *Mad Men*–era 1960s, the Secretary Problem—you consider a predetermined number of candidates one at a time, and if you reject one you can't change your mind later and go back groveling on bended knee. This version of the problem, it turns out, has a precise and provably optimal solution that was worked out a decade after Flood's conference presentation. Whether you're screening suitors or secretaries, you should pass on the first 37 percent of candidates, then choose the next candidate who is better than the best one you've seen so far. Follow this rule, and you'll end up with the best available applicant 37 percent of the time. Call it the 37 percent rule.

Whether it's conscious or not, we generally do pretty well at following this look-then-leap strategy. In a study of job-hiring decisions, for example, subjects successfully chose the best applicant 31 percent of the time, not far from the optimal score of 37 percent. The reason for the shortfall? Most people settled too soon, a finding that shows up in other studies too. That doesn't necessarily mean we're getting it wrong. It may just be that the model is too simple to capture what we're really optimizing. Our time has value, for example. Spending an extra week trudging around inspecting apartments in search of one that may or may not be infinitesimally better than the one you saw on the first day of your search probably isn't a win. Also, we don't usually have perfect information. It's not easy to judge whether the twentieth apartment you tour is marginally better or worse than the fifth one was. Assuming a little bit of uncertainty in your information turns out to nudge you toward stopping early.

The cost of continuing to explore can be psychological too. I often end up with a long walk from parking spot to destination, and as we walk my passengers are happy to point out all the wonderful spots we could have had if we'd kept driving a little longer. The math on optimal parking is more complicated than the 37 percent rule, because you have to factor in what percentage of parking spots are currently occupied. But it's obvious without any calculations that I stop early. I enjoy walking and I hate parallel parking under pressure, so if you add anxiety and mental anguish into the equation, my choice is no longer quite so irrational.

The constraints of the real world make optimal stopping problems challenging even for the greatest minds in history—like Johannes Kepler, the brilliant seventeenth-century German astronomer and mathematician who figured out how planets orbit around the sun. After the death of his first wife in 1611, Kepler famously spent two years sequentially considering the merits and deficiencies of various candidates for remarriage. The fifth candidate, twenty-four-year-old Susanna Reuttinger, tempted him with "love, humble loyalty, economy of household, diligence, and the love she gave the stepchildren." But she was an orphan with no dowry, so Kepler opted to return instead to the fourth candidate. He had waited too long, and she declined his proposal—a complication not considered in Flood's original Fiancé Problem. Eventually, after considering a total of eleven candidates, he decided that Reuttinger was the best candidate after all. As the fifth candidate, she was the one he'd have chosen by sampling the first 37 percent of the eventual pool then grabbing the next good option. "If Kepler had been aware of the theory of the secretary problem," UCLA math professor Thomas Ferguson points out in a historical overview of the field, "he could have saved himself a lot of time and trouble."

Kepler and Reuttinger, by all accounts, ended up living happily ever after. Still, researchers tell this story not to lionize the 37 percent rule, but to point out all the ways that real life deviates from the assumptions necessary to produce such precise guidance. The one thing all solutions to all versions of the Fiancé Problem have is that they begin with a period of pure exploration. That's the takeaway to remember. As for Sue Flood, it's perhaps notable that her official minutes from the conference in 1950 don't contain any mention of her father's Fiancé Problem. Perhaps she didn't want to hear it. "I have frequently used the algorithm for the Fiancé Problem in my personal decisions of various kinds, in a crude form," Flood wrote to a friend several decades later, "and have also inflicted the principle on friends and family members." In this case, though, his daughter didn't need math. When she returned home to California after the conference, she found that the older man was now dating one of her friends.

Nils van der Poel's training manifesto made headlines around the world, in both mainstream and niche outlets. "He Broke Every Record. Then He Told His Rivals How to Beat Him," the *Wall Street Journal* reported; *Trail Runner magazine* lauded his "Wildly Cool Training Approach." Over time it became clear that athletes really were beginning to follow the path that Van der Poel had charted—and not just speed skaters. Andreas Almgren, who swept up every Swedish running record between 1,500 meters and the half-marathon in 2022 and 2023, attributed his newfound success in part to adding cycling to his off-season routine, like Van der Poel. "There probably isn't an endurance athlete in Sweden who hasn't had a Nils van der Poel effect," he said. Cross-country skiers too found that the method was "working beyond all expectations," the website ProXCskiing.com reported.

Van der Poel himself, meanwhile, was wrestling with one of Merrill Flood's optimal stopping problems. At what point would the value of another Olympic medal or world record drop below the physical and psychological cost needed to earn it? Even during the Beijing Olympics, between the 5,000 and 10,000 final, he was teasing the possibility of retirement. "I've been trying to quit two times and I haven't succeeded," he told reporters. "I don't think I'll succeed this time either. Maybe." After the 10,000, he upped his odds of retirement to 80 percent. Then he skated in a few more competitions, including the World Allround Speed Skating Championships, in which skaters rack up cumulative points from races at 500, 1,500, 5,000, and 10,000 meters over two days. He won again, despite his relative weakness at the shorter events. Then he once again packed his skates away. This time, he said, it was for good. Unless he changed his mind.

Did Nils van der Poel really discover a whole new way of training? Even he was circumspect about the idea in his interview with Stephen Seiler: "To this day—I mean, it was very successful—I'm not sure that was the *most* successful way of doing it. Perhaps the other way around would also work excellent." In a radio interview with the Swedish public broadcaster, he was philosophical about this unavoidable uncertainty. After all, he admitted, whether you choose to explore or exploit, you'll never know how the other option might have turned out. "As soon as you go in one direction, you also choose to not go in all other directions at the same time," he explained. "It's the basic precondition for going anywhere at all."

Months passed, and still he didn't skate. He went on the speaking circuit and formed partnerships with companies like GKN Aerospace. Pichler, the Swedish Olympic Committee's talent czar, was reassuring when questioned by the press. "If he wants, Nils could get in shape for the next Olympics if he starts training in two years," he said. But Pichler, for one, wasn't putting any pressure on him. There's a time to explore . . . and there's also a time to exploit. "He barely had any money before he broke through," Pichler said. "I think that Nils should reap the benefits after the inhuman training he submitted to for such a long time, and be sure to make money—and lots of it."

5

UNKNOWN UNKNOWNS

After yet another sixteen-hour day of canoeing, Alexander Mackenzie decided to pull an all-nighter. He and his companions had covered some eighty-five miles since setting out at four o'clock that morning on the thirty-seventh day of their voyage, paddling and portaging through a rugged and increasingly barren landscape. The trees that lined the river's six-foot-high banks—mostly stunted firs and the occasional birch—kept tumbling into the current as the exposed soil thawed in the bright northern sun. Mackenzie planned to stay up to see what time the sun would finally set, in order to determine exactly where he was.

He had left Fort Chipewyan, a newly established fur-trading post on Lake Athabasca about a thousand miles north of present-day Montana, on June 3, 1789. This was about as far west across the continent as Europeans had made it overland at that point. Just a few years earlier, in 1785, the details of Captain James Cook's third (and ultimately fatal) maritime voyage of exploration had been published, including his mapping of the west coast of North America from Oregon up past British Columbia and Alaska and through the Bering Strait. Cook carried with him a chronometer, an innovative new device for keeping accurate time, which enabled him to calculate for the first time the longitude of the west coast. He noted, for example, what appeared to be the outlet of a huge river on the coast of what is now Alaska at 152 degrees west of Greenwich.

Cook's findings were of great interest to the fur traders whose pursuit of beaver pelts was, at the time, the main impetus for inland exploration from the east coast of North America. Mackenzie was a partner in a fur-trading firm based in Montreal, on the St. Lawrence River. The furs he traded for with Indigenous hunters around Lake Athabasca, at the extreme edge of the trading network, had to travel nearly three thousand miles by canoe to get back to Montreal before being shipped back to Europe. The round-trip voyage meant a delay of more than two years before the traders would get paid, so the idea of shortening the voyage by sending furs directly to the west coast was enticing. Just north of Lake Athabasca was an even bigger lake, which we now know as Great Slave Lake. From its western edge, which the fur traders had charted at 135 degrees west, a huge river flowed westward to points unknown. It was only natural to connect the dots: the river flowing west from Great Slave Lake must be "Cook's River" that emptied into the Pacific, perhaps as few as six days' travel away.

Mackenzie set out to test this theory, accompanied by four French Canadian voyageurs, a German ex-mercenary named Johann Steinbruck whose presence deep in the interior of North America remains a bit of a mystery, and a Chipewyan chief named Nestabeck and his wives; two Chipewyan hunters followed in a separate canoe. The territory they paddled through wasn't unpopulated; they regularly encountered groups of Indigenous peoples whose dialects were intelligible to the Chipewyans, at least during the initial stages of the voyage. But none of the people they met could give them any information on the river's ultimate destination; each group only knew (or was only willing to reveal) what lay within a few days' travel. What little information they did share was not encouraging.

"Suffice it to say," Mackenzie wrote in his journal after one such encounter, "that they would wish us to believe that we would be several Winters getting to the Sea, and that we all should be old Men by the time we would return." Not only that: they would encounter "many Monsters" and several impassible waterfalls and rapids.

Mackenzie didn't believe the warnings. But after more than five weeks of hard travel, he had to admit that the trip wasn't going as planned. For one thing, the river kept veering north instead of continuing west. Unlike Cook, Mackenzie didn't have a chronometer or other state-of-the-art navigational equipment with him, so his options for determining his exact location were limited. The rest of the party slept as he waited for the sun to dip below the horizon. And waited. Finally, at half past midnight, he concluded that it wasn't going to set at all. He woke up one of the voyageurs to share the discovery: "on seeing the sun so high, he thought it was a signal to embark, and began to call the rest of his companions, who would scarcely be persuaded by me, that the sun had not descended nearer to the horizon, and that it was now but a short time past midnight." Finally, Mackenzie convinced everyone to go back to bed until 3:45 a.m., at which point they started paddling once more. The river, he noted, was "very serpenting."

The midnight sun was a cool sight, but it was also a reality check: they were north of the Arctic Circle. As they paddled onward, they came across abandoned Inuit camps with whalebones scattered around the campfires, along with the skulls of an unknown animal that he figured must be a "sea horse"—presumably what we now call walruses. They were now perilously close to the point where it would be impossible to make it back upstream to Fort Chipewyan before the waterways froze up again, and Mackenzie was increasingly suspicious that the great river he'd been following led not to the Pacific but to what he called the Hyperborean Sea—that is, the Arctic. But he couldn't quite bring himself to turn around. He decided to press on to the end of the river, "as it would satisfy Peoples Curiosity tho' not their Intentions."

The multi-armed bandit paradigm is a handy way of clarifying the distinction between exploring and exploiting, and of revealing the "uncertainty bonus" that guides our decisions. But what, exactly, do we mean by *uncertainty*? In the bandit scenario, we're weighing probabilities against one another in a casino. Exploration, we might conclude, is the decision to venture a few chips on a long shot in the hopes of a disproportionately big payoff. Uncertainty, in this picture, is simply another word for risk.

But that's a highly unsatisfying way of thinking about exploration. Flipping a coin or rolling a pair of dice isn't exploring. Neither, for a comparison that's a little closer to Mackenzie, is paddling over a waterfall. The world record in the latter discipline is 189 feet, set by a twenty-two-year-old named Tyler Bradt when he kayaked over Palouse Falls in Washington in 2009. That's twenty-two feet higher than Niagara Falls. You could argue that this feat involved venturing into new territory. "It's kind of an unknown realm a bit for kayaking and what the human body can take off of a waterfall," one of Bradt's crew members tells the filmmakers documenting Bradt's exploit shortly before he paddles off toward the abyss. And it's undeniably risky; a couple of years later, Bradt broke his back when another waterfall descent went wrong. Bradt's odds of success might well have been lower than Mackenzie's when he set off for the Pacific, but the potential payoffs were very different. Bradt was seeking a thrill; Mackenzie, in venturing off the edges of the mapped world, was embracing a qualitatively different kind of uncertainty—one that, thanks to the free energy principle, we're wired to recognize as a signal of the possibility of gaining new knowledge about the world.

The most famous—or perhaps infamous—modern exegesis of uncertainty comes from former defense secretary Donald Rumsfeld. In a Pentagon news conference in February 2002, five months after 9/11, he faced questions about the evidence, or lack thereof, that Iraq was concealing weapons of mass destruction in Iraq. "Reports that say that something hasn't happened are always interesting to me," he replied, "because as we know, there are known knowns; there are things we know we know. We also know there are known unknowns; that is to say we know there are some things we do not know. But there are also unknown unknowns: the ones we don't know we don't know."

Whether Rumsfeld was dispensing pearls of wisdom or simply obfuscating, his framework has a long history. Both John Keats and Robert Browning used the phrase "known unknown" in poems in the nineteenth century. In the 1920s and 1930s, the economists John Maynard Keynes and Frank Knight wrestled with the differences between uncertainties with known probabilities and those where the odds themselves are unclear—unknown unknowns, so to speak. "Knightian uncertainty," in economics, has come to mean the degree to which the probability of an event is not just unknown but fundamentally unknowable. The line of research that Rumsfeld was tapping into came a few decades later, when Cold War theorists were debating the finer points of nuclear deterrence—and it started with a young graduate student who Richard Nixon later labeled "the most dangerous man in America."

Daniel Ellsberg is well known as the former military analyst who in 1971 leaked the Pentagon Papers, a seven-thousand-page report detailing the shifting rationales and blatant deceptions that had kept the United States mired in the Vietnam War. He avoided a potential jail sentence of up to 115 years on espionage charges in part because Nixon's Watergate burglars broke into his psychiatrist's office in search of incriminating information, leading the judge to declare a mistrial. The leaked papers had enormous and lasting implications on debates about foreign policy, military intervention, and press freedom. And they vividly illustrated the real-world importance of the somewhat arcane field that Ellsberg had first studied as a doctoral student and fellow at Harvard in the late 1950s: how we make decisions in the face of uncertainty.

Ellsberg's dissertation focused on the distinction between two types of uncertainty, which he called risk and ambiguity. Risk, in Ellsberg's framework, is simply an uncertainty that can be represented by a numerical probability. Calling heads in a coin toss is a risk with a fifty-fifty chance of paying off. Leaving your umbrella at home when there's a 20 percent chance of rain is a risk. Kayaking off a waterfall is, for an experienced extreme kayaker, a risk—not because it's dangerous, which is a different meaning of the word 'risk,' but because based on your own prior experiences and those of others, you can formulate an estimate of the likelihood that you'll be able to stick the landing and emerge unharmed.

Ambiguity, on the other hand, is an uncertainty that you can't quantify. Keynes, in his 1921 book *A Treatise on Probability*, used the bringing-your-umbrella-on-a-walk example. He didn't have weather apps to consult, so he had to integrate various sources of information that were sometimes incompatible: "If the barometer is high but the clouds are black, it is not always rational that one should prevail over the other in our minds, or even that we should balance them." What are the odds of rain in that scenario? Keynes couldn't put a number on it, because his information was ambiguous. More generally, Ellsberg described ambiguity as "subjective uncertainty when experience was lacking, or information was sparse, the bearing of evidence was unclear, the testimony of observers or experts was greatly in conflict, or the implications of different types of evidence were contradictory." Another good example: following an uncharted river into unmapped territory, with no knowledge of the endpoint other than secondhand reports of a river emptying into the Pacific hundreds of miles away.

In a 1961 paper, Ellsberg proposed a set of thought experiments to distinguish risk from ambiguity. One involves two urns, each filled with a hundred colored balls. The first has fifty red balls and fifty black balls; the second has an unknown mix of red and black balls. If offered the opportunity to bet on whether a red ball will be drawn from one of the urns, most people choose to bet on the first urn with the known odds. This implies that they think there are fewer red balls than black balls in the second urn. But if you ask them to bet instead on whether a black ball will be drawn, they'll still opt for the first urn—which implies, in contradiction of their previous bet, that they think there are fewer *black* balls in the second urn. This is what's now known as the Ellsberg paradox: when choosing between two otherwise equivalent propositions, most people prefer the one with quantifiable risk rather than ambiguity.

This pattern, which has come to be labeled "ambiguity aversion," is remarkably widespread and persistent. In a 2023 study, researchers at Yale offered volunteers a series of bets with real money, carefully arranged so that the best odds would be obtained by overcoming any aversion to ambiguity. As expected, the bettors fell prey to the Ellsberg paradox and made suboptimal bets. Then the researchers taught them about the Ellsberg paradox, helped them calculate the correct odds, and explained the optimal betting strategy. That improved their performance in a second round of betting but still didn't eliminate ambiguity aversion entirely. Indeed, brain imaging studies suggest that risk and ambiguity are processed by separate neural circuitry, so it's not surprising that we have trouble weighing them against each other.

After finishing at Harvard, Ellsberg became an analyst with RAND, a nonprofit organization that at the time mostly produced classified research for the air force. As an expert on decision-making and uncertainty, he delved into the intricacies of nuclear control: under what circumstances would the president or his proxies authorize the launch of nuclear warheads? As Ellsberg put it, "It would be the transcendent, and conceivably the last, decision under uncertainty ever made by a national leader." It was also a decision that would inevitably have to be taken under conditions of extreme ambiguity. Shortly after Ellsberg joined RAND, the North American Aerospace Defense Command, burrowed inside Cheyenne Mountain in Colorado, hit the maximum threat level of 5, meaning a 99.9 percent certainty that the country was under nuclear attack. It turned out that signals from a newly commissioned radar complex in Greenland were bouncing off the moon as it rose over Norway. Even more worrying, radio communications in the presatellite era were routinely disrupted by atmospheric conditions. A nuclear bomber squadron scrambled in response to a vague but ominous warning might reach their rendezvous point only to be greeted by complete radio silence—a situation that could mean either that the weather was bad or that nuclear war was underway and their airbase had already been wiped out. What would the pilots choose to do in the face of this ambiguity?

A few years later, in 1964, Ellsberg moved to a Pentagon job focusing on the emerging conflict in Vietnam. His very first day on the job coincided with an apparent attack by North Vietnamese patrol boats on two U.S. destroyers, the USS *Maddox* and the USS *Turner Joy*, patrolling in the Tonkin Gulf. This was the incident that finally drew the United States into active conflict in Vietnam, and Ellsberg experienced it in real time from cables arriving one after the other from Captain John Herrick: "Torpedoes missed. Another fired at us.

And five torpedoes in the water." They had sunk one of the North Vietnamese boats. At least, they thought they had; it was the middle of the night. Then, a few hours later, Herrick reconsidered. "Freak weather effects on radar and overeager sonarmen may have accounted for many reports," he wired. "No actual visual sightings by *Maddox*." It was, in a phrase later associated with Ellsberg's boss at the time, Secretary of Defense Robert McNamara, the fog of war. But by then the wheels were already in motion, and by the end of that day U.S. planes were en route to bomb North Vietnamese targets.

Ellsberg's inside view of how ambiguity affected military decision-making eventually led him to leak the Pentagon Papers. Others, like Donald Rumsfeld, drew different lessons from this line of research: even if we can't quantify the odds that the Soviets have more missiles than us, or that Iraq has weapons of mass destruction, we can't be *sure* that they don't. One way or another, we have to wrestle with the unknown unknowns—the blank spots on our maps. But the concept of ambiguity as a distinct form of uncertainty, and our well-established and in many ways justified aversion to it, raise a challenge for accounts of exploratory behavior. How can we not only be wired to explore and be drawn to "uncertainty bonuses," but also averse to ambiguity?

Two days after staying up to see the midnight sun, Mackenzie and his men reached the entrance to a lake. It was a strange sort of lake. They camped on an island and climbed to a high point, from which they could see mountains, islands, and a solid mass of ice stretching as far as the eye could see. Over the next few days, they found polar bear bones, chased whales—"It was, indeed, a very wild and unreflecting enterprise, and it was a very fortunate circumstance that we failed in our attempt to overtake them, as a stroke from the tail of one of these enormous fish would have dashed the canoe to pieces"—and woke up in the middle of the night to find that the water had risen enough to soak their baggage. The "lake" was in fact the Arctic Ocean, tides and all. The river they had followed for more than a thousand miles is now known as the Mackenzie, boasting the second-largest drainage basin

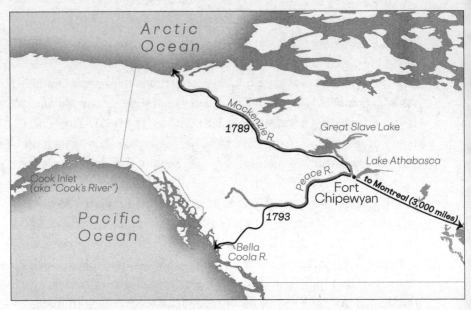

Alexander Mackenzie's Expeditions

in North America after the Mississippi. But Mackenzie dubbed it the River Disappointment. He had thought he was paddling to the Pacific.

If the voyage to the Arctic was a hard slog, the return trip was a frantic race against dwindling food supplies and impending winter, paddling, poling, and lining upstream against the formidable current, all the while "much tormented by Musquittoes." Still, at every opportunity Mackenzie stopped to quiz the Indigenous people he met along the river. Was there, perchance, any *other* river in the vicinity that might take him west rather than north? One group of Sahtú hunters knew of a river that led west, but it could only be reached by crossing the Rocky Mountains on foot. The people who lived at its mouth reportedly made very large canoes and hunted "a kind of large Beaver the Skin of which is almost Red"—sea otters—but were "Big and very wicked" and "kill Common Men with their Eyes." Over the following days, he quizzed other groups, who confirmed that there was indeed a river beyond the mountains, and even drew him a map. They also agreed that the inhabitants were big and killed people with their eyes, and added that they had wings but couldn't fly.

Some of this information, like the existence of a river just over the mountains, was absolutely correct. Some of it was clearly nonsense—possibly, Mackenzie speculated, invented to try to convince him to go away. The information he got from his European contemporaries was no more reliable. The two key data points that had convinced him to embark on his expedition in the first place, it turns out, were both wrong. The mouth of the great river that Cook had reported on the Pacific Coast wasn't a river at all; it's now known as Cook Inlet, in Alaska, rather than Cook's River. And the coordinates of Mackenzie's starting point were way off: rather than 135 degrees west, the western edge of Great Slave Lake is actually at 117 degrees, 565 miles farther inland than he thought. Had he realized how far he was from the coast, he probably wouldn't have set out in the first place.

On September 12, 102 days after their departure, Mackenzie and his companions made it back to Fort Chipewyan. They were just in time: the weather had turned a few nights earlier, with deep frost and snow during the day. Other than reaching the wrong ocean, the expedition had been a remarkable success, traversing more than three thousand miles of harsh and uncharted terrain with no lives lost and no shots fired in anger. But Mackenzie knew he needed better information. "In this voyage, I was not only without the necessary books and instruments, but also felt myself deficient in the sciences of astronomy and navigation," he later wrote. "I did not hesitate, therefore, to undertake a winter's voyage . . . in order to procure the one and acquire the other." He needed tools like a telescope powerful enough to observe Jupiter's moons, along with the knowledge to use them, and he spent the winter of 1791 and 1792 in London getting them. He returned to Fort Chipewyan in 1792, and in May 1793 he set off once again, this time with the goal of crossing the Rocky Mountains and finding a new river that would take him to the Pacific.

The metanarrative of the modern-day adventurer often goes something like this: I noticed a great but neglected injustice in the world; I wondered how I could raise awareness about this injustice; I decided to undertake this unrelated but epic journey to attract the attention of the world and raise money to remedy the injustice. We all understand that the actual chain of logic is usually the other way around: you start with a desire to go on the voyage and reverse-engineer a way to justify it. This isn't a new phenomenon, George Loewenstein points out: "many expeditions that seek to accomplish 'firsts' disguise their true purpose by including a 'scientific' or 'humanitarian' component." Even more than a century ago, Ernest Shackleton's ill-fated Antarctic crossing required "the blessings of the government and of various scientific societies . . . to justify the expedition as a serious scientific endeavor."

This tendency makes it tricky to disentangle Mackenzie's true motivations from what might simply have been convenient justifications. Here's what he wrote in the introduction to his expedition journals, which he eventually published in 1801:

> I was led, at an early period of life, by commercial views, to the country North-West of Lake Superior, in North America, and being endowed by Nature with an inquisitive mind and enterprising spirit; possessing also a constitution and frame of body equal to the most arduous undertakings, and being familiar with toilsome exertions in the prosecution of mercantile pursuits, I not only contemplated the practicability of penetrating across the continent of America, but was confident in the qualifications, as I was animated by the desire, to undertake the perilous enterprise.

In the next paragraph, he describes his exploration as "this favourite project of my own ambition," but also boasts it "extends the boundaries of geographic science, and adds new countries to the realms of British commerce." Based on his personal correspondence and public actions in the years following his voyages, he really does seem to have been extraordinarily eager to expand his company's fur-trading network and extend British sovereignty to the Pacific. But he also persisted along the Mackenzie River even after it became clear that continuing "would satisfy Peoples Curiosity tho' not their Intentions." In other words, he's got the full mix of Daniel Berlyne's proposed motivations for exploration: internal predisposing factors, pursuit of rewards and biological utility, and ludic behavior.

In the simplest picture of our explore-exploit algorithms, we might claim that Mackenzie was drawn onward by an uncertainty bonus. There were, at that time, beaver pelts available in pretty much every direction from Fort Chipewyan. He knew roughly what was available to the north, south, and east—but the potential rewards of going west, particularly the prospect of a faster way of shipping his pelts to international markets, were unknown and thus more attractive. On the other hand, what information Mackenzie did have was as ambiguous as any garbled wartime cable from Vietnam—or, for that matter, as the distribution of colored balls hidden in an urn. True to Ellsberg's definition, experience was lacking, information was sparse, and the testimony of observers was greatly in conflict. Ambiguity aversion should have made a westward voyage highly unattractive. Of course, that aversion might be overcome by the prospect of gaining valuable information that would earn him big piles of money. How do all these pushes and pulls interact?

Robert Wilson, a computational cognitive scientist at Georgia Tech, confronted a version of this question in the early 2010s, when he was a young postdoctoral researcher at Princeton. He was studying explore-exploit strategies in humans, trying to figure out whether we use uncertainty-directed exploration, random exploration, or a mixture of both. Previous attempts to answer this question had produced conflicting results, and Wilson thought he knew why: the design of typical multi-armed bandit experiments mixed ambiguity and reward in a way that made it impossible to disentangle them.

Imagine you're playing a two-armed bandit game. Initially, both arms are maximally ambiguous: you know nothing about their respective risk-reward profiles. Then you start playing, and after a few pulls, one arm will have produced better results than the other arm, so you'll tend to keep pulling the more rewarding one. The more you pull it, the less ambiguous it gets, which means that higher reward and lower ambiguity become correlated with each other. The same thing happens in the real world: we tend to pursue rewarding options more often than unrewarding ones, so those high-reward choices also tend to have lower ambiguity. But if we want to understand how our brains weigh these different factors, we need to find a way of separating them.

Wilson designed a modified two-armed bandit protocol, which he dubbed the Horizon Task, to tease apart the respective influences of ambiguity, uncertainty, and reward. The protocol has two key features. The first is that, instead of letting you choose which arm to pull, the game starts by forcing you to pick one arm or the other for the first few rounds. In this way, Wilson can either ensure that you have the same amount of information about each arm, meaning that they're equally ambiguous, or can provide more information about one arm than the other. The second feature is that he manipulates the time horizon. Sometimes you get only one turn, which means that there's no benefit to gaining information about the properties of the two bandits because you won't get to use that information in future turns. Other times you get six turns, which makes gaining information from exploration more valuable.

By manipulating these variables, Wilson can zero in on which factors affect the likelihood that you make an exploratory choice—that is, opt for the arm with the lower reward (based on the information you currently have) in order to gain more information about it. Sure enough, Wilson's first Horizon Task results, published in 2014, found that when ambiguity is held constant, people really do engage in more uncertainty-directed exploration when there's something to gain from acquiring information. And they make seemingly "wrong" decisions more frequently in the six-turn version of the game, showing that they also ramp up random exploration when there are benefits to exploring.

Daniel Ellsberg was interested in the distinction between two kinds of uncertainty, which he called risk and ambiguity. But Wilson's Horizon Task, by teasing them apart, reveals that neither risk nor ambiguity is really what we're seeking when we explore. The free energy principle suggests that the goal of exploration is information that will help you minimize future surprise. Not all information is useful, though. Tossing a coin has an uncertain outcome, but knowing whether it came up heads or tails doesn't teach you anything about what the result of future tosses will be—unless, University of Waterloo experimental psychologist Elizabeth Lapidow points out, you suspect it's a trick coin and you're gathering data about its properties. Similarly, learning about the properties of a bandit arm in the Horizon Task isn't useful in the one-turn version of the game. "The drive to explore is a drive to learn," Lapidow says. "So uncertainty isn't the underlying driver; potential to learn is—but uncertainty is often present where that potential is."

In other words, what we've been calling an uncertainty bonus should really be called an information bonus (which is, in fact, what Wilson calls it in his Horizon Task paper). In practice, as Lapidow points out, uncertainty is often a good proxy for the potential to learn. If you're a rat in maze, or a fur trader choosing which river to explore, you can't predict a priori which path will lead you farthest toward your goal or reveal the most about your surroundings. All you can do is go in the direction of your greatest uncertainty and hope that it turns out to be informative—that the maze doesn't hit a quick dead end, and the river doesn't bend to the north.

In 2023, Lapidow and Elizabeth Bonawitz published a study comparing the effects of ambiguity, reward, and information gain on explore-exploit choices in preschoolers. Which of these factors has the strongest influence on our exploring decisions? In a series of experiments, the kids had to choose between two boxes in order to get as many marbles as possible, which they could then use to play in a marble maze. The "exploit" choice was a box that they knew contained two marbles. The "explore" choice was a box with an unknown number of marbles, selected at random from among a group of boxes whose properties were manipulated to control the levels of ambiguity, reward, and information. In effect, the study directly pitted the exploration-dampening effects of ambiguity aversion against the lure of an uncertainty—or information—bonus.

In the first experiment, the kids knew nothing about how many marbles the unknown boxes contained: the explore choice was completely ambiguous. In a tug-of-war between the lure of uncertainty and aversion to ambiguity, the latter was stronger: the kids chose to explore, on average, just 36 percent of the time.

The second experiment tested quantifiable uncertainty—what Ellsberg called risk—instead of complete ambiguity. Once again, the kids could "exploit" by choosing a box with two marbles, or "explore" by choosing from a set of unknown boxes. But this time they knew the *average* number of marbles in each set of unknown boxes, which ranged from half a marble to as many as eight marbles per box. As you'd expect, they were more likely to explore when the potential rewards were highest, opting for the unknown boxes 81 percent of the time when the expected payout was eight marbles and just 32 percent of the time when the payout was half a marble. But the most telling scenario was the middle one: when the average number of marbles in the unknown boxes was two, meaning that the expected value of the explore-exploit decision was a fifty-fifty toss-up, the kids explored 67 percent of the time. That's consistent with the idea that we—or at least children—are biased toward exploring quantifiable uncertainty beyond what the odds suggest is rational.

The third and final experiment introduced the possibility of information gain. The children started with no knowledge of how many marbles the unknown boxes contained: complete ambiguity. But when they chose to explore, they were told that the number of marbles in their selected box was a clue to how many marbles were in other boxes that they could choose from in subsequent rounds of the game. As a result, exploring provided information that could be useful for future decision-making. In this scenario, the kids explored 77 percent of the time, even though they initially faced the same ambiguity that suppressed exploration in the first experiment. In this particular scenario, at least, the lure of useful information overpowered ambiguity aversion, which in turn was stronger than the draw of uncertainty on its own. And it's tempting to see the same forces at work in Mackenzie: the call of the blank spots on the map, the doubt about what might await there, and the irresistible prospect of gaining information that would make him rich and famous.

Mackenzie's 1793 voyage was, despite his improved navigational competence, another trip into the unknown. His plan this time was to head west—and upstream—on the Peace River, which flowed from the Rocky Mountains into Lake Athabasca near Fort Chipewyan. Then he would cross the Rockies on foot and look for a river that would take him west to the Pacific Coast. That's more or less how it worked out, though not without some missteps.

The first blunder was that he somehow missed an incoming river, now known as the Pack River, that would have given him an easier route over the mountains. This is the route that the explorer Simon Fraser found in 1806, just over a decade later—and Fraser was incredulous that Mackenzie hadn't even mentioned this "large and navigable" river or included it on his maps. "As he used to indulge himself sometimes with a little sleep, likely he did not see it," Fraser wrote. And indeed, Mackenzie admitted as much. In fact, just the day before he would have passed the Pack, he had lost his notebook: "I was in the habit of sometimes indulging myself with a short doze in the canoe," he wrote, "and I imagine that the branches of the trees brushed my book from me, when I was in such a situation."

Eventually, with the help of a map drawn with charcoal on a strip of bark by a man from the Sekani First Nation, they made it to the Continental Divide. There they found a well-worn portage trail just 817 paces long—about seven football fields—between two small lakes that took them from the Arctic watershed to the Pacific watershed. Then they followed a series of waterways—James Creek (which Mackenzie dubbed Bad River after they smashed their canoe and nearly drowned in a set of violent rapids), Herrick Creek, the MacGregor River, and eventually the majestic Fraser River. They followed the Fraser for several days before deciding that its current and rapids made it all but impassable by canoe and that, even if they could follow it, it was leading too far south rather than west. After quizzing more groups of Indigenous people, Mackenzie decided to retrace his steps up the Fraser and, over the strenuous objections of his voyageurs, abandon his canoe and strike out overland toward the coast.

It turned out to be a good decision. After crossing a snow-covered pass through the Coast Mountains at six thousand feet above sea level, they found the Bella Coola River, where they borrowed two canoes from the hospitable inhabitants of a Nuxalk settlement that Mackenzie called Friendly Village. The Bella Coola River, at long last, took Mackenzie all the way to the shores of the Pacific Ocean. He spent a few days exploring the labyrinth of inlets and islands along the coast, then prepared to head back. But first he mixed some vermilion with melted grease, and on the face of a rock overlooking the sea, inscribed these words:

Alexander Mackenzie
from Canada, by land,
the twenty-second of July,
one thousand seven hundred and ninety-three.

The aftermath of Mackenzie's voyage was underwhelming. He'd been disappointed that few people seemed interested after his 1789 trip to the Arctic: "My expedition is hardly spoken of," he complained in a letter to his cousin. Now he'd made it to the Pacific, covering 1,200 miles on the outbound trip in seventy-four days and making the return journey in less than a month. But the route he'd blazed across the Rockies wasn't very practical for trade purposes. When he finally returned to Fort Chipewyan, he plunged into a depression. He spent that winter manning the trading post, hoping to use the time to prepare his journals for publication, but found he was incapable of getting anything done. When spring came in 1794, he traveled back east to Montreal and began his transition from frontiersman to merchant. He was around thirty years old, and he never went west or explored again.

It took another seven years before Mackenzie finally managed to publish his journals, with the help of a ghostwriter, in 1801. The swashbuckling tales of two trips into the unknown were popular with readers and boosted his reputation. He was knighted in 1802; Thomas Jefferson read a copy that same year, sparking his plans for an American expedition to the Pacific. When Lewis and Clark completed that trip—the *second* crossing of the continent north of Mexico, despite frequent claims to the contrary—they were almost certainly carrying a copy of Mackenzie's book with them. After squabbling with his fur-trade partners and penning endless letters to the British government urging them to set up a trading empire on the Pacific coast of North America, Mackenzie eventually got married, retired to an estate in Scotland, and died eight years later, at the age of fifty-six, of what was likely kidney disease.

Maybe this is what a happy ending looks like. Christopher Columbus, after his famous voyage of discovery, made three more trips across the ocean, each progressively more disastrous than the last, eventually costing him his reputation, his fortune, and his health. John Franklin, who in 1825 became just the second European to make it to the mouth of the Mackenzie River, was still at it two decades later, at the age of fifty-nine. He died along with the other 128 members of his 1845 expedition to the Northwest Passage after their ships got jammed in the ice. Was Mackenzie crazy to give up the excitement of frontier exploration for the humdrum routine of the merchant, or was he wise to avoid Franklin's fate? Is exploring a young person's game?

6

THE TIME HORIZON

In a limestone cave about eight miles east of Tel Aviv, among the rich trove of prehistoric artifacts left from hundreds of thousands of years of human habitation, are a bunch of really lousy stone tools. Some of the flint cores have long, smooth faces where sharp blades have been "knapped" off with a hard blow. But others are pocked with irregular scars that produced nothing useful. "The beginners, they try again and again, and repeat the same mistakes," says Ella Assaf Shpayer, one of the archaeologists from Tel Aviv University who excavated the site. "And they get mad, so you even see their frustration in the core. It's a very emotional thing, knapping."

These botched tools—as frustrating as they were for the would-be toolmakers—reveal intimate details about how ideas spread in the Lower Paleolithic period. Some were clearly the work of complete beginners, while others had a curious mix of expert and novice strokes. Shpayer believes that they illustrate a broader pattern that has likely recurred throughout human history, and may be one of our species' most powerful innovations: at key turning points, when new ideas and technologies emerge in response to social or environmental change, it's the children who lead the way.

In this particular case, the trigger for change was likely the disappearance of elephants from the Near East, which took place around four hundred thousand years ago. Forced to hunt smaller and faster animals like fallow deer instead, the inhabitants of Qesem Cave got smarter and more agile, developed more sophisticated tools, and built a large central hearth in their cave. "Fire has a cultural meaning,"

Shpayer says. "It makes everybody come and sit together. Then you see people talk and convey messages. They share experiences with one another and share their food. Fire took the whole learning process to the next level that we don't see in primates." It's around the hearth that experts knapped high-quality blades and then handed over the flint cores to children watching eagerly over their shoulders, to let them have a try.

Making a stone blade isn't easy, as Shpayer herself discovered when she attended a workshop with an expert knapper in Spain during her doctoral studies. "Look," she tells her students when she shows them the blades she made, "this is how it looks when you do a bad job." So how did early humans master the required skills and spread them across the region? The accumulation of cultural knowledge is one of humankind's key superpowers, and it relies on a complex mix of social and cognitive developments—including, Shpayer says, the admittedly maddening way that your toddler figures out the advanced features of your new phone before you do. Kids, the evidence suggests, are wired to explore.

In 2019, the psychologists Jean Twenge and Heejung Park published an analysis of survey results from eight million American teenagers between 1976 and 2016. Across a wide range of milestones of adulthood—working for pay, driving, going out without parents, having sex, dating, drinking alcohol—the results suggested that kids these days are taking longer to grow up. This kind of observation, of course, has been around for a long time. "Longer Adolescence Is Held Key to Rise in Venereal Ills," a 1971 *New York Times* headline proclaimed (though the rise in gonorrhea and syphilis, a Department of Public Health official acknowledged, could also be related to a "national preoccupation with sex"). And there's no doubt that current notions of childhood in the developed world have changed a lot since the Dickensian conditions of the Industrial Revolution. Charles Dickens himself began working in a boot-blacking factory when he was eleven or twelve, after his father was thrown into debtors' prison.

But hand-wringing over the purported failure to launch of today's young people misses a broader point: compared to other species, we've *always* had a stunningly long childhood, one so out of whack with those of even our closest relatives that it demands explanation. Chickens are born with the prewired ability to move around, feed themselves, and flee from predators without any instruction from their parents. Horses can get up and walk within an hour of birth. By the time they're around seven, even chimpanzees can gather as much food as they eat. Human hunter-gatherers, in contrast, typically don't produce as much as they consume until their mid to late teens—not *that* different from Gen Z.

The life history of any organism is shaped by its environment. That's why animals that live in dangerous settings tend to reproduce early and often, for instance. What factors, then, gave us such a long time under the parental wing? Archaeological evidence suggests that long childhoods emerged during the same period that other hallmarks of modern humans did: bigger brains, tool use, language, cultural transmission, and perhaps a roving disposition. It could be simply that bigger brains need more time to develop. But in a series of papers in the 2010s, Berkeley cognitive psychologist Alison Gopnik advanced a more utilitarian hypothesis: our long childhoods, she argued, are an evolutionary solution to the explore-exploit dilemma.

The basic logic is easy to follow. "There has always been an intuition that kids are explorers and old folks are exploiters," says Robert Wilson, the Georgia Tech cognitive neuroscientist. "And in many ways it makes sense that that would be your life trajectory. You want to learn things while you're little and then you want to make the most of them as you get older." But Gopnik takes it a step further. Many of the traits we associate with young children—they're "noisy, variable, unfocused, unpredictable and impulsive," she writes—are in fact features rather than bugs. They help children explore the world around them with remarkable efficiency, and in some cases they pick up clues that adults miss.

In one study, Gopnik and her colleagues tested 106 preschoolers and 170 college students on their ability to figure out which of a set of ceramic blocks were "blickets." Putting one or more blocks on a "blicket detector" would make the box light up and play music if and only if a blicket was detected. In some cases, individual blocks triggered the machine; in others, only certain *combinations* of blocks did. In the straightforward condition when a block either was or wasn't a blicket, children and adults were equally good at picking up the pattern. But in the more complex case where only a combination of blocks triggered the machine, the children were significantly better at picking up this unexpected rule.

A follow-up study tested how susceptible kids and adults are to falling into "learning traps" from not exploring enough. This time, the subjects had to figure out which wooden blocks were "zaffs," with rewards for correct guesses and penalties for incorrect ones. Zaffness depended on both the color *and* the pattern of stripes or spots on the blocks. After an incorrect guess, adults tended to assume that they knew the rule—that all black blocks weren't zaffs, say. In reality, only *spotted* black zaffs weren't zaffs, but the adults never learned this rule—they fell into the learning trap—because they stopped exploring black blocks entirely. Kids, in contrast, kept exploring and were more likely to figure out the correct two-part rule.

There's a catch, though. The kids didn't actually outperform the adults overall, because even after they'd learned the correct rule, they sometimes still chose to put an incorrect block on the zaff detector, seemingly just for the fun of seeing what would happen. In a sense, that's further evidence of the developmental trade-off Gopnik proposes: kids are better at exploring, but adults are better at exploiting. That doesn't mean the kids are doing something wrong, points out Charley Wu, who heads the Human and Machine Cognition Lab at the University of Tübingen. "Children are not optimized for the silly little games that we make them play in the lab," he says. "Evolution has crafted them to be very, very good at learning about the world more generally. And so in the very narrow confines of our little experiments, they may be over-exploring. But in terms of what their life goal is—to learn about the world and to become good decision makers—it pays off."

In an explicit test of Gopnik's ideas, Wu's student Anna Giron led a 2023 study that analyzed performance in a multi-armed bandit task in almost three hundred participants ranging in age between five and fifty-five. As expected, the kids did more exploring and the adults were better at exploiting. But it wasn't just a question of making more random and unpredictable choices. Younger subjects also did more uncertainty-guided exploration, preferentially zeroing in on choices with the greatest potential to gain new information. And as they got older, they gradually improved their ability to generalize from their existing knowledge to make better future choices.

The particular combination of these three ingredients—random exploration, uncertainty-guided exploration, and generalization—dictated how well any given participant performed in the bandit task. The kids tended to be all over the map in how they mixed these ingredients, albeit with an overall bias toward exploration. But the adults were far more similar to one another. We don't just explore less as we age; rather, we seem to converge on an increasingly optimal mix of exploring, generalizing, and exploiting. In fact, when Giron and her colleagues ran computer simulations to come up with ideal parameters for their bandit task, none of the simulations outperformed the human adults—suggesting, they write, "a remarkable efficiency of human development."

The price of that efficiency is uniformity. Confronted with a given problem, most of the adults in the room suggest similar solutions—solutions that are, on average, better than the harebrained schemes your six-year-old might suggest. But if, instead of a stable and predictable world, you inhabit a rapidly changing environment (or if you're a volunteer in one of Alison Gopnik's blicket experiments), you don't necessarily want everyone to come up with the same old solutions to a challenging new problem. You want diversity. You want novel and unexpected suggestions. You want to explore like a child.

Partway through her doctoral research on the cultural transmission of toolmaking, Ella Shpayer gave birth to twins. "It really made me think a lot about the ways we teach, the ways we learn, and about how, in our Western society, we always try to tell kids what to do," she says. Gopnik's thesis is that the goal of childhood exploration isn't just to optimize your individual life trajectory, but also to help societies adapt to change. The ancient humans who fashioned stone tools in Qesem Cave were facing a period of rapid upheaval after roughly a million years of relative stability—but the evidence suggests they did a good job of harnessing that superpower. "They gave their children a lot of freedom to experience and do things by themselves, and you can see it in the cave," she says. "They allowed their children to explore, and make mistakes, and try again."

From a purely mathematical perspective, the case for exploring when you're young then exploiting when you're old is clear. More generally, anytime there's a clock running, exploration is most valuable when you've got plenty of time left, and declines in value as time runs down. In a multi-armed bandit game, you should explore more when you've got a hundred turns left than when you've got ten turns remaining; in college, you should spend more time making new friends as a freshman than as a senior. Time-limited explore-exploit problems are actually easier to solve than the open-ended ones that John Gittins eventually tackled. A mathematician at the RAND Corporation named Richard Bellman worked out the details in the 1950s, and his equations can be used to show that, for example, you should gradually reduce the share of risky—that is, exploratory—assets in your portfolio as you approach the "endpoint" of retirement.

We seem to be pretty good at following this rule. You can see it in action on two very different timescales in Robert Wilson's Horizon Task. As we saw in the last chapter, the Horizon Task involves a multi-armed bandit game where you get either one or six turns. Subjects reliably explore more at the beginning of the six-turn game than in the single-shot game but get gradually less exploratory as they proceed through their six turns. That's exactly what the math predicts you should do. Now zoom way out: when Wilson and his colleagues compared the Horizon Task performance of people in their sixties and seventies to that of younger adults between eighteen and twenty-five, the older subjects explored less overall than the younger ones. That too is exactly what the math predicts you'll do, since on average you've got less time remaining on the planet when you're older. But is it really what you *should* do?

I can't help seeing an analogy between exploring and being physically active. Both decline as we age, for reasons that are easy to understand. But over the last few decades, there's been a shift among physiologists and exercise scientists away from a fatalistic acceptance of the body's deterioration. We don't get more sedentary simply because our bodies begin to fail us; our bodies deteriorate because, as adults with steadily mounting family and career pressures, we allow ourselves to become less active. A 2022 analysis by researchers in Switzerland concluded that about half the typical losses in fitness that accompany aging are the result of changes in exercise habits rather than of aging itself.

Is the same true of exploring? Or to put it another way, how much of the decline in exploring through the lifespan is hardwired into our brains, and how much is a response to our changing circumstances? If you had the time and freedom you had as a kid, would you regain your itch to explore? Would you *want* to?

There are no clear answers to these questions, but there are some clues to consider. One is that the adult brain is physically different from the child's. We start out with a hypernetworked brain featuring an overabundance of connections between neurons. As the brain matures, and as we learn which patterns of brain activity are most useful, those connections are trimmed back on a "use it or lose it" basis. By the time we reach maturity, our brains have about half as many synaptic connections between neurons. The very architecture of our brains, in other words, guides our thoughts along a progressively smaller network of more well-traveled paths.

Even if you deliberately choose to explore as much as possible, you still won't quite be able to recapture what you were like as a kid. The reason, according to Charley Wu: "It's hard for us to regrow our expectations about the world." Wu—whose academic path so far has taken him from Toronto to Vancouver to Vienna to Berlin to Boston to Tübingen—gives the example of moving to a new city. Initially you have very broad expectations, and new experiences beckon from every door. If you've lived somewhere for a long time, on the other hand, you already know the patterns and rhythms of local life. "Some new restaurant pops up, and right away you're like, 'Oh, that's the hipster coffee shop,'" he says. "You haven't even gone in. It's not even open yet, but you already know what to expect." That, writ large, is what adulthood is like: for better or worse, we already know whether we like hipster coffee—and if we don't, then there's no joy in exploring the new place.

There's another way of looking at the link between time horizon and exploratory behavior. If time horizon dictates behavior, then observing behavior should give us some clues about time horizon. In their 2016 book *Algorithms to Live By*, Brian Christian and Tom Griffiths illustrate this point by looking at the prevalence of blockbuster movie sequels. Creating a brand-new movie is a huge and expensive risk; revisiting a previously successful franchise is a textbook exploit. In 1981, they point out, just two of the ten highest-grossing films were sequels. In 1991, it was three. In 2001, it was five. In 2011 it was eight. "By entering an almost purely exploit-focused phase," they write, "the film industry seems to be signaling that it is near the end of its interval."

The trajectory hasn't changed since then. In 2022, *all ten* of the highest-grossing Hollywood hits were sequels or (in the case of seventh-ranked *The Batman*) reboots. And it's not just a case of what audiences choose to watch: sequels have taken over a steadily growing share of the overall output of Hollywood studios. Without original new movies, where are the sequels of tomorrow going to come from? Judging from their actions, the studios don't think it matters. Barry Diller, the former chairman and CEO of both Paramount Pictures and 20th Century Fox, came right out and said it in 2021: "The movie business as before is finished, and will never come back." If that's the outlook, then of course you squeeze every dollar out of the surest bets you've got.

That type of analysis casts a different light on Twenge and Park's data on longer childhoods. If today's children are spending longer in the exploratory phase of development, perhaps it's because they're preparing for a longer lifespan. The life expectancy for babies born in the United States increased by about ten years between 1950 and 2010, so it's perfectly rational that Generation Z might choose to spend a few extra years trying out different options and searching for an ideal fit—in work, love, and life—before settling in for the long exploit.

In addition to a longer time horizon, greater instability might also favor the extension of exploratory childhoods. The optimal policy in a restless bandit game involves more exploration than in a stationary one where the rewards stay fixed. It's pretty clear that the modern world is more restless than, say, Qesem Cave was four hundred thousand years ago. The fossil record in the region shows remarkable cultural stability over periods of hundreds of thousands of years, and even innovations like new ways of making stone blades emerged over many generations. Since then, a series of revolutions—agricultural, industrial, technological—have revved up the pace of societal change dramatically. In the early 2000s, young workers wrestled with the realization that—unlike their parents—they probably wouldn't end up spending their entire career with one company. These days, journalists like me wonder whether our entire industry will even exist in another decade—and we're far from alone, with the continued advance of industrial automation and the rapid rise of artificial intelligence.

So in this respect too spending a little extra time exploring your options and acquiring diverse skills is more likely to pay off these days than it might have been even a few generations ago. It makes sense that millennials born between 1980 and 1996 are three times more likely to report having changed jobs within the past year than older generations. As someone who moved back in with my parents at twenty-four, and abandoned my first career to start a journalism degree at twenty-eight, I'm happy to say that this extended pseudochildhood worked out well for me. I love being a journalist, and I think I'm better at it for having studied science first. But it's not just about me. In Gopnik's developmental framework, a society that gives its young plenty of opportunity to explore works better for everyone, particularly during periods when tomorrow will look a lot different from today.

When I was twelve years old, I started playing the alto saxophone in band class at school. My parents, supportive as always, trudged off to a record store and asked the salesclerk for the best saxophone music to inspire young prodigies. Under the Christmas tree that year, I found the soundtrack to *Bird*, the Clint Eastwood–produced Charlie Parker biopic that had debuted earlier that year.

Parker, who died of a heart attack in 1955 at age thirty-four after years of drug abuse, was the soaring id of the bebop movement in jazz. In contrast to the staid and often formulaic big-band swing popular in the 1930s and 1940s, bebop was harmonically and rhythmically complex, prizing dissonance and unpredictability, and demanding extreme technical virtuosity of its practitioners. Louis Armstrong, avatar of the old guard, dismissed bebop as "Chinese music" and bemoaned its "weird chords which don't mean nothing." Worst of all, he added, "You got no melody to remember and no beat to dance to."

That was pretty much my impression when I placed my new record on the turntable that Christmas morning. I couldn't even tell the album's eleven tracks apart: to my ears, it was just a wall of undifferentiated noise.

This memory bubbled up for me, thirty-five years later, when I was interviewing Marc Malmdorf Andersen, the Danish cognitive scientist we met in chapter 3. Andersen's theory of play sees it as a counterpart of exploration: slope-building versus slope-chasing. In both cases, we're drawn to intermediate levels of complexity and predictability, because that's where the most fruitful opportunities to reduce uncertainty lie. That's true of music too. At a fundamental level, we're slope-chasing, making a stream of predictions that are sometimes fulfilled and sometimes confounded. Music that is too predictable is boring; but music that *always* lands awkwardly on a note you don't expect feels unpleasant. Studies have found that our musical preferences follow a Wundt Curve, with the greatest pleasure elicited when melodies, harmonies, and rhythms are neither too predictable nor too unpredictable. But what counts as predictable depends on your prior experiences. "That's why musicians listen to jazz music and the rest of us think it sounds like shit," Andersen told me. "Some of us want to listen to AC/DC. Jazz is too surprising for us, because we haven't played as much music as the musician has. 'Highway to Hell' is plenty surprising to me, right?"

That's exactly how I felt about the *Bird* soundtrack. Still, it was the only album I owned, so I'd occasionally put it on to marvel at the blizzards of notes that Parker could produce with his horn, or just for an excuse to play with the record player needle. Over time, I began to be able to tell the tracks apart. There was one called "Now's the Time" that had a relatively rhythmic and repetitive melody. (Not coincidentally, its melody was subsequently lifted for a successful R&B dance hit called "The Hucklebuck.") At the other extreme, a tune called "Ko Ko" still sounded to me like a deliberate attempt to make the most random and obnoxious noise possible. I remember once playing it for a visiting friend purely for shock value, to show him what depravities humans were capable of.

In spirit, I would have been right at home had I attended the most infamous premiere in music history, the 1913 debut of Igor Stravinsky's *The Rite of Spring* in Paris. Right from its opening notes, the dissonant and seemingly arhythmic ballet sparked laughter, then catcalls—"Call a doctor!" a listener begged. "A dentist!" another amended—then vegetable-tossing and full-throated jeers that drowned out the music. There was almost a riot. The critic from *La Figaro* called it "laborious and puerile barbarity." Another critic made a more astute observation: "The music always goes to the note next to the one you expect."

Predictions evolve, though. As famous as the *Rite of Spring*'s opening night is, the reception to subsequent performances is often forgotten. "At the second, there was noise only during the latter part of the ballet; at the third, 'vigorous applause' and little protest," the music critic Alex Ross recounts. "At a concert performance of *Rite* one year later, 'unprecedented exaltation' and a 'fever of adoration' swept over the crowd, and admirers mobbed Stravinsky in the street afterward, in a riot of delight."

As you've no doubt predicted, Parker's blizzards of sound ended up growing on me. I started to get interested in jazz, which led me to start listening more carefully. Three years after I got the album, my jazz quintet performed publicly for the first time at a school music night. We played "Now's the Time." And by the time I finished high school, my favorite track on the album was "Ko Ko." By then I understood the harmonic logic underlying the tune, and I'd learned to play a transcription of Parker's improvised solo. I struggled to understand how I'd ever found it dissonant and chaotic. One of the quotes I chose for my high school yearbook was from a Robertson Davies novel, foretelling what Marc Malmdorf Andersen would tell me decades later: "Music is like wine, Bridgetower. The less people know about it, the sweeter they like it."

That's not the end of the story, though. In the years that followed, my tastes continued to gravitate toward more challenging and esoteric music. I was taking university courses in jazz theory and arranging, soaking up new perspectives on melody, harmony, and rhythm. Even Charlie Parker started to sound . . . predictable. After university, though, the tide shifted. I was wrapped up in my career and devoting most of my free time to running, and moving between various cities in Canada, Britain, the United States, and Australia, which made it hard to find consistent bandmates to play my sax with. I noticed my tastes contracting again. I was listening more to my earliest CD purchases and dropping the more challenging ones from the rotation.

I certainly wasn't alone in this musical regression. There's a rich body of academic research that seeks to understand why people tend to stop exploring new music as adults. Not everyone does, of course. The journalist Tim Falconer, who wrote in the book *Bad Singer* about his attempt to finally learn how to sing as an adult, recounts an awkward moment in the men's room at a concert at Toronto's Kool Haus. "Aren't you a little old to be here?" a soul-patched kid a few urinals away asks him. Falconer was in his forties at the time, as eager as ever to keep discovering new favorite bands. But people of his generation were—and are—mostly nonexistent when touring bands played clubs in town, other than when aging rockers came through on stadium tours.

By and large, then, our preferences are locked in by the time we reach adulthood. Various studies have pegged the age of peak musical exploration at between seventeen and twenty-three years old, and the songs we discover in those years tend to stay with us for the rest of our lives. Our music trajectories, in other words, follow the explore-then-exploit developmental framework laid out in Alison Gopnik's research. As kids, we're wired to explore widely and discover the music that turns our cranks. As adults, we're wired to sit back and enjoy the resulting playlist for the rest of our lives. "Most people have all the songs they could ever need by the time they turn thirty," Jeremy Larson, director of reviews for *Pitchfork*, wrote in a 2020 essay. "Why spend time on something you might not like?"

Larson, of course, believes that listening to new music as an adult is "hard, but necessary." You're helping artists and pushing the boundaries of art, he suggests, and perhaps creating the soundtrack through which a future you will remember the current moment. That lines up with my own intuition about the importance of continuing to explore, both for music specifically and for life in general. Andersen, though, is leery of suggestions that adults should be pressured to play more. The whole point of his slope-chasing and slope-building framework is that we do it because it feels good, not because we ought to. "If a person becomes interested in something, and they have time, and they don't have all these other fires to put out, then they'll follow this hedonistic principle all by themselves," he says.

Those are some big ifs: *if* you have time, *if* you don't have other fires to put out, *if* the endless constraints and responsibilities of adult life just fade away. Charley Wu too doesn't necessarily think adults should mindlessly insist on exploring rather than exploiting. "If you're in this Peter Pan–style mode of never getting old," he says, "then you never really settle into your niche." But as a busy academic in his thirties launching a research career, he is also conscious of the growing constraints on his ability to chase slopes. "I feel like I was more future-oriented around the time I started grad school," he admitted toward the end of our conversation. When he moved to Berlin for his PhD, for example, he invested huge swaths of time in learning German. "I had all these things I wanted to do, because I had a horizon for exploration. Now it sometimes feels like I'm just trying to survive to the next day, and making sure I'm prepared for my next meeting or lecture.... I have so many things on my plate that it's hard to imagine thinking about, say, learning a new language."

If you follow the time horizon argument to its logical conclusion, there will eventually come a time when you should stop exploring entirely—preferably a few moments before your heart stops beating for the final time, though it's tricky to nail the timing precisely. Robert Wilson's Horizon Task data showed that, on average, older people explore less than younger people. But there's a worrying signal when you look at the individual data rather than the overall averages: some older people seem to be getting the timing very wrong.

As with all behaviors, there's plenty of variability within groups. "There are some young people who don't explore very much," says Wilson, "and there are some old folks who keep exploring." But what jumps out in the data is that roughly a quarter of the older adults essentially never explore at all. They almost always choose the option that they know more about, even when the other option is clearly the better and more informative choice. Wilson dubs this behavior "extreme ambiguity aversion." Most psychology experiments produce data that is smoothly distributed on a continuum, but these nonexplorers are way out at the edge of the distribution, seemingly playing by a different set of rules.

Like most of the researchers in this field whom I've spoken to, Wilson is cautious about making normative pronouncements about how we "should" behave. His research focuses on observing what kinds of exploring decisions people make naturally, and understanding how and why we reach those decisions. But the subset of extreme nonexplorers concerns him: "I think a certain amount of reduced risk-seeking is good as you age, but I suspect completely avoiding it is pathological." You don't want to become like J. R. R. Tolkien's hobbits, who view adventures as "nasty uncomfortable things" that might make you late for dinner, he says. You'd eventually reach a point where you stop engaging with the world and never leave your home.

Extreme ambiguity aversion is very rare, though not completely unheard of, in younger adults. Over the years, Wilson and his colleague have seen it in about one percent of the more than five hundred subjects they've tested. There's just one exception: in a 2020 study of the exploring patterns of patients with schizophrenia, 21 percent of them showed extreme aversion. For now, it's unclear whether there are any links between the nonexploring psychiatric patients and the corresponding older adults, who were screened to exclude anyone with psychiatric conditions or signs of early cognitive impairment that might have affected the results.

The question is whether the near-complete disappearance of exploration in a quarter of people is a sign that something has gone wrong—some sort of change in brain structure or decline in dopamine signaling, say. Or perhaps it's simply the normal endpoint in an evolutionary life history where the benefits of exploration decline with age. But if so, it's not clear that this template still serves us well. "I don't want to get too evolutionary psychology about it," Wilson says, "but it's clear that we now age in very different ways." Instead of staying with family, modern seniors often move to a retirement community, or even to a different city, where they might live for several decades, meeting new people and taking up new hobbies, all while adjusting to ever-changing technologies. As a result, the mindset that enables a grandparent to thrive in the twenty-first century is probably substantially different from what worked in Qesem Cave.

There is one final insight from Wilson's study—an encouraging one. The nonexplorers, not surprisingly, performed worse overall in the bandit task, meaning that they earned a lower total return on their bets. But if you take them out of the analysis, the remaining sexagenarians and septuagenarians actually outperformed their younger peers, even though they explored less. This suggests that, at least among the three-quarters of people who don't stop exploring entirely, we're continuing to refine and improve our instincts. So it's okay if, in your sixties, you're no longer quite as eager as you were in your teens to see every new place and try every new flavor. That's a rational response to the effect of time horizon on explore-exploit decision-making. But you don't want to neglect that itch entirely. There's evidence, Wilson points out, that exploring new interests and ideas—quilting and digital photography were the examples used in one randomized study—helps protect against cognitive decline. And you never know how much time you'll have to reap the benefits of exploration: "If you're sixty years old," he says, "your average time horizon could still be twenty years, right? Which is quite a long time."

7

MAPPING THE WORLD

Robert O'Hara Burke was the first to spot them, peering into the darkening twilight as he stumbled along the banks of Cooper Creek, in the barren interior of Australia. "I think I see their tents ahead!" he exclaimed.

Burke had set out from Melbourne in August 1860, leading an expedition that initially consisted of twenty-five men, six wagons, twenty-six camels, around forty horses, and four dogs. Like Alexander Mackenzie in North America, his goal was to complete the first crossing of the continent—but instead of cold that stabbed like a driven nail, his primary foes were heat, thirst, and hunger.

Cooper Creek was roughly the halfway point, and on December 16, 1860, Burke and three others left the main expedition party and most of its supplies there to make a final push to the northern coast, with just six camels and a horse to carry supplies. Burke asked the base-camp party to wait at Cooper Creek for three months for their return. But his second-in-command, William Wills, had a private word with William Brahé, who was in charge of the base camp, on the morning of their departure. Stay for four months, if at all possible, he requested.

Burke's advance party successfully reached the coast, achieving their goal of crossing the continent. It was a brutally hard trek, though, and progress was slow. Now, as they staggered back into the Cooper Creek base camp on April 21, 1861, four months and five days after their departure, they were at their limits. One of the men, Charles Gray, had died four days earlier. He'd been suffering from dysentery since killing and eating a python, and then Burke had given him a beating after he was caught stealing more than his share of their dwindling gruel rations. They had long since eaten their horse and four of the camels. They were barely able to move, Wills wrote in his journal, "our legs almost paralyzed, so that each of us found it a most trying task only to walk a few yards." Instead, for the first time in their journey, the three survivors were now riding the two remaining camels, Burke on one, Wills and the other member of the party, John King, on the other. When they finally reached the campsite that evening, a fire was still burning and equipment was scattered around. But Burke's eyes had deceived him in the autumn moonlight: there were no tents. "I suppose they have shifted to some other part of the creek," he said uneasily.

Then Wills noticed blazes on two coolabah trees, sap still running from the gashes. One of them read: *B LXV DEC 6–60 APR 21–61*. It was the sixty-fifth camp of the Victorian Exploring Expedition, known to posterity as the Burke and Wills expedition, with the arrival and departure dates of the base-camp party. Wills turned to the others. "They have left here today," he said. On the other tree was a single word in all caps: *DIG*.

Much of our discussion so far has treated exploration as an abstract concept. Life, in this metaphor, is an imaginary casino where every decision can be framed as a choice between slot machines with different payoffs. To explore is to pull a lever with a lower but more uncertain reward. But this way of thinking doesn't really capture what it's like to explore in the physical world. What does the distinction between "random" and "uncertainty-directed" exploration look like when, say, you're standing on the banks of Cooper Creek, contemplating possible routes to the northern coast through the surrounding labyrinth of dunes, mesas, and desiccated creek beds?

The way we explore new places is shaped by the way our brains encode the landscape around us. To decide where to go next, you need to have some concept of where you've come from and where you currently are. You have a mental map of the world as it's known to you, and exploring is the process of adding to that map. That means figuring out the spatial relationships between the new things you encounter and the landmarks and paths you already know. If, when visiting a new city, you call an Uber, plug in an address, then sit in the back seat and stare at your phone until you arrive at the destination, you haven't explored the city. You have no idea where you are relative to your hotel, how to get back there, and what sorts of neighborhoods and attractions lie between the two places.

You can put these abstract maps—scientists call them cognitive maps—into a more concrete form by sketching them on a piece of paper. In the late 1960s, a Berkeley psychologist who was helping to plan the redesign of Ciudad Guayana in Venezuela showed that car drivers in the city could draw more detailed and accurate maps than people who generally traveled as passengers on buses or in taxis. Subsequent researchers have confirmed that how you travel—and particularly, whether you're an active navigator or a passive passenger—has a strong influence on the cognitive maps you form. Kids who walk to school, for example, can draw far more detailed, accurate, and feature-rich maps of their neighborhoods than those who are driven.

There's a chicken-and-egg evolutionary question here: Did we learn to form cognitive maps because we were intrepid explorers, or was it cognitive maps that made us intrepid explorers in the first place? Either way, our cognitive mapping abilities are intimately connected to our propensity to explore. But their significance extends beyond crossing untracked deserts or finding your way back to your hotel in an unfamiliar city—though in the desert, as Burke and Wills discovered, the details of a cognitive map can spell the difference between life and death. They also shape how we understand and explore the landscape of ideas, a topic we'll delve into in the next chapter. We'll consider too what it means to live in a technology-enhanced world where, for better or worse, it's increasingly possible to live without cognitive maps.

In 1901, an experimental psychologist named Willard Small, of Clark University in Massachusetts, published the results of a study inspired by the famous hedge maze at Hampton Court Palace in London. Using wood and wire mesh, he constructed a miniature six-foot-by-eight-foot version of the maze, with an open area—the goal—in the middle. Over the months that followed, he ran a series of experiments with both wild and tame rats, timing how long it took them to reach food in the middle of the maze, counting their errors when they turned down blind passages, and observing their general behavior—which, initially at least, didn't strike him as particularly impressive. The rats seemed to make small forays into the maze, then double back, then head out again in a different direction before returning, Clark wrote, "back and forth, digging at the base of the wall and biting at the wires—the depth of stupidity, one would say."

Still, the rats would reliably find the food eventually. And once they'd found it the first time, they were quicker to find it in subsequent attempts. They had learned something, in other words—though what exactly that learning consisted of remains a topic of ongoing debate. Did they simply memorize a series of turns and landmarks? Or did they somehow encode a more global sense of where the food was? If so, how did they orient themselves? Clark's study was the first maze study in psychology, but far from the last. "The history of psychology," according to behavioral neuroscientist Paul Dudchenko, "is, in part, a history of how rats find their way in mazes."

The idea that we carry around cognitive maps in our minds dates back to a 1948 rats-in-mazes paper by Edward Tolman, a professor of psychology at the University of California, Berkeley. At the time, John Watson and B. F. Skinner's behaviorist theories were ascendant in psychology: the actions of both lab rats and humans, Skinner argued, could be understood as a straightforward matter of stimulus and response. Put rats in one end of a maze and food at the other end, and the rats will soon learn to navigate efficiently through the maze to get to the food as quickly as possible. They're not intrinsically motivated to explore, or curious about what lies beyond the next corner. They just want the tasty pellet that they've learned awaits them after they follow a rote pattern of lefts and rights.

Tolman, in contrast, didn't buy what he saw as a reductive view of the mind as "a complicated telephone switchboard." In "Cognitive Maps in Rats and Men," his 1948 paper, he laid out the case for a more nuanced mental representation of space in our minds. There were prior hints even in the behaviorist literature, he pointed out. During a series of maze studies in the 1920s, a behaviorist named Karl Lashley had removed the maze's mesh lid to see what his trained rats would do. Some of them hopped onto the roof of the maze and made a beeline directly for the finish—a navigational feat that would have been impossible if they were just following a sequence of left and right turns. Tolman's most famous experiment confirmed this accidental finding. He trained rats to find food at the end of a maze, then altered the maze's configuration. The rats were able to adjust and pick the most direct route to their reward.

This ability to find shortcuts is the key distinction between memorizing a route and mapping a landscape. To Tolman, an idealist who at one point lost his Berkeley job for refusing to sign a McCarthy-esque loyalty oath, a good cognitive map was as essential for understanding the true relationship between ideas as it was for plotting a course from A to B. His 1948 paper concludes with a passionate plea for "the child-trainers and world-planners of the future" to foster the development of broad cognitive maps to promote harmony among races, religions, and sexes in "that great God-given maze which is our human world." This is lofty stuff for the staid pages of the *Psychological Review*, and a bit of a leap from the rats-in-a-maze data it's based on, so perhaps it's not surprising that the concept of cognitive maps languished in obscurity for a few decades after Tolman proposed it. Besides, even those who found the idea interesting had no way of peering into the brain and searching for this supposed map—yet.

The eventual breakthrough started with an accident. In a lab at University College London in 1970, an American researcher named John O'Keefe was trying to record the activity of single neurons in rats' brains, focusing on a region called the somatosensory thalamus. But in one experiment, he missed and ended up implanting a microelectrode in the rat's hippocampus, a seahorse-shaped structure that at the time was mostly thought to be involved in long-term memory.

He was immediately struck by the pattern of the neuron's firing, which seemed to be linked to the rat's movements, and decided to conduct further experiments on hippocampal neurons.

"And then on one electrifying day I realized with a flash of insight that the cells were responding to the animal's location or place in the environment," he recalled years later, when he was awarded the 2014 Nobel Prize for his epiphany. Each individual neuron fired when—and only when—the rat visited a specific location in its room. These "place cells," as he dubbed them, formed a literal map of the rat's physical environment, and an embodiment of Tolman's cognitive map—which, O'Keefe noted, was "a vague hypothetical construct that [Tolman] had used to explain some aspects of rodent maze behavior but which had never gained much acceptance in the animal learning field."

In the years since then, scientists have identified other components of our neural navigation system: "grid cells" that mark our exact coordinates in a space (discovered by the Norwegian pair May-Britt and Edvard Moser, who shared O'Keefe's Nobel), compass-like "head direction cells" that encode which way we're facing, and "boundary cells" that keep track of how far we are from borders or edges in our physical space. The borders are important, because our place cells are like an Etch A Sketch, wiped clean each time we enter a new space. Where one cognitive map ends and another begins depends on the context: simply passing through a doorway is a common transition point. Indeed, one of the clues that we use cognitive maps to organize our thoughts more generally is the "doorway effect," the all-too-familiar (and laboratory-validated) feeling of walking from one room to the next only to realize that you've suddenly forgotten what you came for. Such insights suggest another definition of exploring: it's what happens when you step outside the borders of an existing cognitive map and start forming a new one.

In 1606, a Dutch navigator named Willem Janszoon, sailing in the service of the Dutch East India Company, became the first European to reach Australia. He mapped about two hundred miles of the northeastern coastline, mistakenly assuming it to be an extension of previously discovered New Guinea, and had a few brief and bloody onshore encounters with the Aboriginal residents of the region. Over the next two centuries, various Dutch, French, and British voyagers—some on deliberate missions to explore, others blown off-course from trade routes—pieced together the outline of Australia. Permanent European settlement followed a similar pattern, starting in 1788 with a British penal colony near the site of present-day Sydney and followed by other coastal outposts in Melbourne, Adelaide, Perth, and Brisbane—all still on the continent's perimeter.

There's a word for this pattern of exploration: *thigmotaxis*. Its literal meaning refers to an organism's response to touch, but in the context of exploration it has come to mean sticking close to the edges or boundaries of a space as you reconnoiter it. Thigmotaxis is ubiquitous in the animal world: cockroaches scurry along baseboards; deer trails run along fences and tree lines; humans cluster along the edges of parks and public squares. Even on curb-less pedestrian-only streets, the urbanist Jane Jacobs observed, most walkers still move along the sides rather than the middle unless it's crowded.

There's an obvious evolutionary explanation for thigmotaxis: it's easier to run and hide from a predator if you're foraging along the edges of the forest rather than out in the middle of the open savanna. Anxious mice spend more time along the edges of mazes than bolder mice, which supports the idea that thigmotaxis is linked to ancient fear responses. The same is true in humans: in one study, volunteers with "high anxiety sensitivity" were more likely than a control group to avoid passing through the open center of a town square during a navigation task, even though it was the most direct route.

But there's another reason we like to stick close to boundaries: they help orient us. Walk into a new space—a room, a neighborhood, a forest—and the place cells in your hippocampus will carve up the area into distinct "fields," with each cell firing if and only if you wander into its corresponding field. The geometry of these fields is determined by the boundaries of the space. In the 1990s, John O'Keefe and a colleague showed that moving one wall to stretch a square room into a rectangular room caused rats' place cell fields to stretch from circular blobs to elongated ovals. This finding led O'Keefe to predict the existence of specialized cells whose sole purpose is to register your distance from spatial boundaries—a prediction that was finally proven to be true in 2009 when boundary cells were identified in rat brains.

So the coastal—or, if you prefer, thigmotactic—pattern of early European exploration in Australia follows the usual template, albeit on a much vaster scale than normal. Once permanent settlements were established, the colonists inevitably wanted to know what lurked in the interior. But their progress was hindered by the Great Dividing Range on the east coast, and by the unremittingly harsh and arid conditions they found farther inland. Despite a series of intrepid and occasionally fatal exploring expeditions, most of the interior remained blank on European maps well into the second half of the nineteenth century—or worse, the maps showed a massive but entirely mythical Great Inland Sea where the rivers of the continent supposedly drained.

By 1860, it was increasingly clear that there was no inland sea. Instead, the invention of the telegraph had given new impetus to the still unaccomplished goal of crossing the continent. The best place for an undersea cable to connect Australia to the rest of the world was the sparsely inhabited northern coast, since the already connected Indonesian island of Java was relatively nearby. To reach the continent's main cities, it would then have to cross the unmapped interior. Colonists in South Australia thought the overland telegraph route should connect to its capital city, Adelaide; those in Victoria, five hundred miles east, thought it should go instead to Melbourne. One version of the Burke and Wills story, then, is that their expedition was spurred by the desire to blaze a potential telegraph route to Melbourne before a rival expedition from Adelaide, led by John McDouall Stuart, beat them to the punch.

There's some truth to this: the Royal Society of Victoria's Exploration Committee, which started planning the Burke and Wills expedition in 1857, listed as one of its objectives "to facilitate our intercourse with the other hemisphere." But as you read through the minutes of their interminable meetings, you get the sense that the telegraph was a convenient fig leaf, a post hoc rationalization for what they already wanted to do. They also suggested that inland exploration might locate valuable minerals and new pastureland, and perhaps turn up traces of the explorer Ludwig Leichhardt, who had disappeared in the interior a decade earlier. But the one goal that stayed constant through the three-year planning process was simpler and less practical: "to enlarge the boundaries of knowledge." There was a huge blank spot on the map—nine hundred million acres, by one estimate—and its massive uncertainty bonus was simply too much to resist.

Once you venture away from the boundaries, wayfinding gets more complicated. A classic test of exploratory behavior is the Morris water maze, developed by Richard Morris of the University of St. Andrews. It's basically a circular swimming pool with one small platform hidden just below the surface. In Morris's original 1981 study, the water was mixed with milk to make it opaque and the platform was painted silver, making it essentially invisible until the animal is right on top of it. The task for rats or mice, who hate swimming in cold water, is to explore a seemingly featureless expanse of open water to find the platform where they can take a rest. Thigmotaxis won't help. They have to venture away from the perimeter, into the interior.

Scientists have run literally thousands of studies with the Morris water maze, and they classify the behaviors they observe into eight distinct exploring strategies. The first and most common is thigmotaxis, followed by incursions—the back-and-forth forays into the interior noted by Willard Small back in the early 1900s. Then there are various subtypes of scanning and search behaviors, some more or less random and others systematically targeting certain areas, analogous to the random and directed exploration observed in multi-armed bandit studies. Sometimes the rodents make little 360-degree circles to figure out where they are, a tactic labeled "self-orienting." If you rerun the experiment multiple times, the strategy mix shifts as the animals start using clues they've picked up in previous runs to zero in on the target more quickly.

You can run similar experiments in humans using virtual reality.

Amazingly, you get pretty much the same results. In one study, German researchers pitted eighteen mice in a water maze against eighteen undergrads in an equivalent virtual setup, searching for treasure on a desert island instead for a platform in a swimming pool. The two groups deployed the same set of exploratory strategies, a similar shift in strategies as they learned from repeated trials, and even had comparable times to find the treasure or the platform. It's not a big leap, then, to think of nineteenth-century Australian colonists as subjects in a continental-scale Morris water maze—or rather, its inverse. Would-be explorers weren't searching for a platform in the ocean, or even for a giant inland sea. Instead, once they abandoned thigmotaxis and headed into the arid interior, the challenge was to find rare sources of life-giving water, often invisible from a distance, before you perished from thirst.

The Victorian Exploring Expedition left Melbourne on Monday, August 20, 1860, to the cheers of fifteen thousand well-wishers. Burke, an Irish ex-soldier and policeman, had instructions to proceed north to Cooper Creek, an intermittent watercourse about halfway up the continent that the explorer Charles Sturt had reached in 1845. From there, he was to venture into the "ghastly blank," as the unmapped interior was then known, and try to reach the northern coast, more than two thousand miles away.

The expedition's secret weapon was twenty-six camels specially imported from India for their desert-crossing skills and their ability to march onward without water for days on end. They had plenty to carry. Twenty tons of supplies were loaded into six horse-drawn wagons, including a cedar-topped oak table with matching stools, 160 pairs of pants, and a Chinese gong. Subsequent chroniclers have understandably ridiculed the seemingly delusional outfitting for an arduous desert trek. "Were twelve sets of dandruff brushes and four enema kits really necessary?" the journalist Sarah Murgatroyd asked in her influential 2002 account of the expedition.

The criticism isn't entirely fair. David Phoenix, who retraced the expedition's entire route for his PhD dissertation at James Cook

University, points out that dandruff brushes—what we would now call dandy brushes—are absolutely essential for removing grit and sweat before saddling horses, especially in hot and dusty conditions. And some of the other much-mocked extravagances listed on the expedition's list of stores, like a bathtub and six tons of firewood, were intended for use only while the expedition was assembling in Melbourne. Still, Burke and Wills, an English-born surveyor and surgeon who was in charge of navigation, weren't traveling light. It took them two months to haul their increasingly bedraggled crew to Menindee, the last output of colonial settlement; the progress report that they sent back by the regular mail coach that served the town reached Melbourne in just two weeks. Burke began shedding men, animals, and equipment as he pushed onward toward Cooper Creek.

Beyond Menindee, the would-be explorers hired a succession of Aboriginal guides as the party passed through the territories of the Paakantyi, the Karenggapa, and the Yandruwandha. Wills knew the latitude and longitude of the point on Cooper Creek that Sturt had reached, and unlike Alexander Mackenzie on his first expedition, he knew how to calculate his own position. But he didn't know anything about the intervening territory (Sturt had approached Cooper Creek from the southwest rather than the south), and he didn't have any way of telling the guides where he wanted to go. So they continued heading roughly north, with the guides leading them from waterhole to waterhole, until, roughly four hundred miles north of Menindee, their last guide refused to proceed any farther. For the rest of the way to Cooper Creek, and then onward to the northern coast, they would be venturing into the unknown every day without any knowledge of what awaited them. (They did encounter other groups of Aboriginal people, some of whom were friendly and offered them food, but none was willing to guide them—even when Burke offered one of their camels, an object of great fascination, as payment.)

In a Morris water maze, some mice make minor incursions into the open water, returning to the safety of the boundary each time. Others just go for it, swimming straight across the pool until they either find the platform or bump into the other side. John McDouall Stuart,

Burke's rival who was trying to blaze a south-to-north path across the continent from Adelaide, followed the former approach. Between 1858 and 1860, while the Royal Society of Victoria was fussing over the planning of Burke's expedition, Stuart launched four separate expeditions. He traveled light, at least in comparison to Burke, and relied only on horses. From each camp, scouting parties would scour the terrain until they found suitable water, then return to guide the whole expedition forward to the next camp. With each expedition, Stuart penetrated farther into the interior. By the time Burke finally reached the Cooper, in November 1860, Stuart was already preparing to head back out on his fifth expedition.

Burke, meanwhile, faced a decision when he reached Cooper Creek. He still needed to travel almost a thousand miles due north to the coast, but he had no idea what the intervening terrain would be like, or whether he'd find water on the way. Initially, he tried Stuart's approach of minor incursions. Wills led four separate scouting expeditions from Cooper Creek, venturing as far as sixty miles north and nearly dying when his camels ran off, but he found no water sources. With temperatures hitting 110 degrees Fahrenheit and the height of Australian summer looming, Burke couldn't wait around at the Cooper forever. So he decided to do it the other way. He took four men, a horse, and six camels loaded with eight hundred pints of water, enough to keep the party alive for ten days, and steered north toward the coast in an all-or-nothing gamble.

The first barrier was Sturt's Stony Desert, which had taken a week for Charles Sturt to cross in 1845, and whose sharp stones had made Sturt's horses go lame. But luck was with them. After just two days without water, they noticed three flocks of pigeons passing overhead at sunset, all heading due north. Just as Polynesian navigators watched for birds as a sign of land, Wills figured the birds were heading to water. They immediately repacked their bags and, starting at 7:00 p.m., hiked another nine miles in pursuit of the pigeons. They didn't find water that night, but at daybreak they continued for another two miles and found a creek that hadn't dried up. Despite their "exaggerated anticipation of horrors," as Wills put it in his journal, they had ended up crossing the Stony Desert at an unusually narrow point. They celebrated with a rest day, it being Christmas Eve. "We had never, in our most sanguine moments, anticipated finding such a delightful oasis in the desert," Wills wrote.

A few days later, they found what's now known as the Diamantina River, and followed it for the next four days. But its course veered farther and farther east, and Burke finally decided that they needed to leave the river and head due north again, into the waterless plains. With the ability to pack ten days of water at a time, they had a five-day window before they would have to either find water, turn back, or—if they continued—face the risk of running out of water in the desert. They pushed hard, putting in eleven hours on the first day and thirteen and a half hours on the second day, covering as much as fifty miles a day. Their water bags started leaking, forcing Burke to reduce their ration from seven to five pints per person per day. On the third day, they found a creek where they could reset the clock by refilling their water bags. Then they set off again—and this time, they found water on the fourth day.

They were now approaching the Tropic of Capricorn, and the landscape began to shift from arid desert to lush tropical wetlands. They had no further problems finding water, and Burke's gamble looked to have paid off. With several mountain ranges still ahead, Burke decided to keep heading straight north and go straight over them rather than trying to weave through or around them. It was heavy going, and they had to drive the camels hard, "bleeding, sweating, groaning" as Burke noted in his journal. Finally, on February 11, 1861, they reached the flooded salt flats of the Gulf of Carpentaria on the northern coast. Although the extensive tidal flats, ankle deep in water, prevented them from reaching the open ocean, they had achieved their goal. All that remained was to make it home safely.

As they traveled north, Wills had been charting their progress in his surveyor's field books: their path in a line down the middle of the page, with bearings and distances to the mountains, creek beds, and other landforms they passed carefully plotted on either side. Little by little, he was transforming the "ghastly blank" into a mapped space, with features whose relative positions he could visualize and draw from a bird's-eye perspective. In the process, he was transforming his relationship with the landscape from egocentric (what do I see when I scan the horizon, and where are these things relative to my current position?) to allocentric (where are these things *relative to each other*?).

These two ways of organizing space in your mind correspond to two different ways of finding your way around. In egocentric mode,

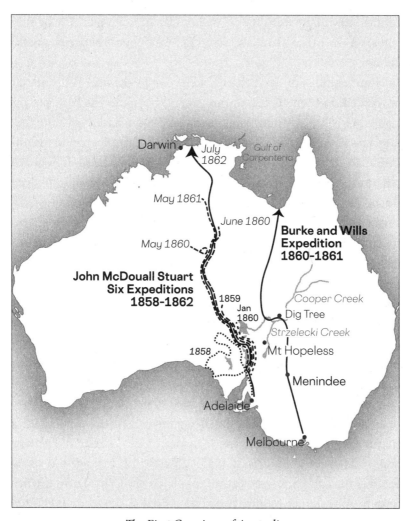

The First Crossings of Australia

you use what researchers call "stimulus-response" navigation. You learn to follow specific routes from point to point: when you see the gas station, you turn right, then go straight until you reach the top of the hill, and so on. In allocentric mode, on the other hand, you use your cognitive map to understand where you are relative to your goal, and what direction you need to proceed in. The big advantages of stimulus-response navigation are that it's faster, less error prone, and takes less mental effort. The big disadvantage is that it's inflexible. You may know how to get from A to B and from A to C—but you have no idea how to get from B to C without backtracking to A. Cognitive maps, on the other hand, allow you to plot a course between any two points.

That flexibility proved to be crucial for Burke and Wills as they proceeded on their thousand-mile journey back to the base camp at Cooper Creek. The monsoonal rains had begun, and they—and their camels—found themselves slogging through knee-deep mud for mile after mile. Eventually the flooding forced them to abandon their outbound tracks and plot an alternate route along the higher (and drier) ridges of the Selwyn Range. Stimulus-response navigation was no longer an option. Their condition was increasingly precarious: they'd consumed three-quarters of their food on the journey north, and the miserable traveling conditions meant they were making even slower progress on the return journey. But Wills, triangulating from the notes he'd scrawled on the grueling trip north, guided them unerringly along the new route back toward Cooper Creek.

The power of cognitive mapping may even have helped propel our ancestors into behavioral modernity, some anthropologists believe. There's a striking pattern in the location of prehistoric archaeological sites in Europe. Neanderthals tended to cluster along the edges of geographical regions—thigmotaxis again—where the topographical features were more distinctive, favoring stimulus-response navigation. Anatomically modern humans, in contrast, managed to spread out over the comparatively featureless plains of eastern Europe, where cognitive mapping would have been required to get around. Hunting and gathering in these areas, as well as maintaining social networks over increasingly long distances, would also have favored the development of cognitive mapping. These demands might have shaped our ancestors' brains, creating a widening cognitive gap between them and their Neanderthal rivals.

The idea that how we navigate shapes our brains isn't far-fetched. Most famously, a 2000 study found that London taxi drivers, with their intricate knowledge of the city's byways, had abnormally large hippocampi. But the effects can be far more subtle. In 2003, researchers at McGill University led by Giuseppe Iaria and Véronique Bohbot tested how volunteers found their way around in a virtual reality world that was specifically designed so that it could be navigated using either stimulus-response navigation or cognitive mapping. In this context, where either approach was reasonable, they found that about half of the volunteers chose each method—and that the difference showed up in their brains. The cognitive mappers were using their hippocampus, as you'd expect from O'Keefe's findings, while the stimulus-response navigators were relying on another region called the caudate nucleus. Amazingly, in an echo of the taxi study, they subsequently found that these brain regions were actually enlarged based on which navigational strategy the subjects preferred.

Both stimulus-response navigation and cognitive mapping are useful in different contexts, but as we go through life we tend to gravitate more toward stimulus-response. Another one of Bohbot's studies found that 84 percent of children used cognitive mapping in her virtual world, compared to 46 percent of young adults and just 39 percent of older adults. It's a transition that mirrors the shift from exploring to exploiting as we get older. Forming a cognitive map, by definition, is an error-prone process that requires exploration; following a stimulus-response route exploits previously collected knowledge. With experience, we get better at this sort of goal-directed navigational exploitation, and so we use it more. In the process, though, our cognitive mapping skills get rusty—with, as we'll see later, potentially worrisome consequences.

Even today, the site of Burke and Wills's base camp on Cooper Creek is impressively remote. "Driving from Brisbane," the conservation area's website explains helpfully, "you take the Adventure Way and drive west for 1,360 kilometers, about twenty hours driving time." The nearest place to get gas, water, or electricity is more than seventy miles away. Still, as many as fifty thousand people a year make the journey and pay the twenty-dollar entry fee. They come to see one of the 250-year-old coolabah trees that Brahé blazed, its slender blueish-green leaves popping out from the washed-out grays of the surrounding outback. To some it's a symbol of intrepid heroism and fortitude; to others, a monument to bumbling incompetence and the overweening hubris of the country's colonial past. It's known as the Dig Tree now.

Once they'd gotten over the disappointment of finding their Cooper Creek campsite deserted, the three surviving explorers followed the engraved instructions and dug—or rather, Wills and John King did. Burke was "too much excited to do anything," King later recalled. Buried at the foot of the tree, they found a box with some provisions and a note. They were too weak and exhausted to follow that night—and after a day of rest, they concluded it would be hopeless to try to catch up with their expedition mates, who were traveling with fresh horses and camels. Neither could they hope to make it more than five hundred barren miles south back to Menindee with limited provisions and only two camels to carry water and food. Instead, they opted to veer southwest in a desperate attempt to reach a remote cattle station near the inauspiciously named Mount Hopeless, which was just 150 miles away.

They reburied the box, leaving some of their journals and a note, without altering the blaze to show that they'd returned. Two weeks later, two members of the relief party made a five-day trek back to Cooper Creek just in case Burke and Wills had somehow managed to return. They found it deserted and seemingly undisturbed, and after fifteen minutes they headed back south. Burke and Wills later returned to the site one more time, too late yet again. It's these repeatedly missed connections, like some lesser Shakespearean comedy, that have earned Burke and Wills immortality. They made some smart decisions, some questionable ones, and some truly incomprehensible ones. They endured incredible privations with fortitude. But in the end, after eight months of wandering, the margin between triumph and tragedy was a matter of hours. Burke and Wills eventually starved to death; King was taken in and fed by the local Yandruwandha people and rescued eight weeks later. The following year, John McDouall Stuart made it from Adelaide to the north coast, near the site of present-day Darwin, on his sixth expedition. The Australian Overland Telegraph Line, built in the early 1870s, followed Stuart's route, as does the modern Stuart Highway. That's where the story usually ends.

But there's a navigational postscript to the Burke-and-Wills story, one that David Phoenix unearthed during the years he spent laboriously retracing the expedition's footsteps. How we explore, after all, is a function of the landscape we're exploring. Burke's seemingly quixotic decision to try to reach Mount Hopeless is usually dismissed as "foolish" and "incomprehensible," perhaps even the result of diminished mental capacity brought on by months of malnutrition and exhaustion. But the route to Mount Hopeless was actually quite straightforward and had been traveled successfully by the explorer and surveyor Augustus Gregory in 1858. They had to follow Cooper Creek until it intersected Strzelecki Creek, then follow the latter south all the way to Mount Hopeless.

In other words, they just had to make one left turn. But they missed it. Watercourses in the Australian outback are famously fickle, meandering and splitting into branches that disappear into the dust and reappear somewhere else entirely. Strzelecki Creek, though it flows south from Cooper Creek, actually gains five or six meters of height over its first three miles, so it flows backward after rainfalls. When Phoenix visited the site, he found a thin trickle of muddy water splitting off the main channel of the Cooper, which soon petered out into a dry creek bed—no different from numerous other braided channels and dead ends along the Cooper. It's no surprise that Burke and Wills missed it. And by the time they'd spent another three weeks trekking back and forth following dead ends in search of the Strzelecki, their food supplies were once again depleted and their final two camels were dead. In Phoenix's view, it's this sequence of events, rather than the more famous Dig Tree debacle, that sealed Burke's and Wills's fate.

Edward Tolman might have chalked up their fate to the perils of following a narrow rather than broad cognitive map. In Véronique Bohbot's terminology, it's the difference between stimulus-response navigation—turn left at the creek—and versatile cognitive mapping. With a few weeks of food and two camels to carry water, they might have succeeded in crossing the desert to Mount Hopeless directly, even without finding the outlet of Strzelecki Creek. But in the desperate rush back from the northern coast, Wills had buried his sextant and lost his copy of Gregory's map. He knew where he wanted to go, and he knew one way to get there. But if you don't have a broader sense of how the various points on your map relate to each other, your explorations are incomplete. And that's equally true for the landscape of ideas, it turns out, where breakthroughs can sometimes seem as elusive as a waterhole in the outback.

8

THE LANDSCAPE OF IDEAS

In the late 1990s, I caught a nasty case of the so-called Oxford flu. I was a grad student at the University of Cambridge at the time, working toward a PhD in condensed matter physics at the hallowed Cavendish Laboratory. It's an inspiring but daunting place to be a young physicist. You're always conscious of the shadow of immortals like Isaac Newton, whose apple tree (or at least a descendant grafted from the original) you walk by every day on Trinity Street. Thirty members of the Cavendish—and counting—have won Nobel Prizes, and along the corridors the muttonchopsed visages of textbook names like James Clerk Maxwell, Lord Rayleigh, J. J. Thomson, and Ernest Rutherford gaze down at you expectantly.

The most daunting part of all, though, is the fundamental requirement for attaining a doctorate at the university: "the creation and interpretation of new knowledge." I'd earned my place at Cambridge by excelling at solving the riddles posed by my undergraduate professors. A single question in the weekly problem sets might require three or four pages of dense mathematical manipulations. But we always knew that the answer existed, and that the route to the answer could be found somewhere in that week's lecture notes. We were traversing terrain that had already been exhaustively mapped. To create new knowledge, on the other hand, would require a plunge into the unknown—"Voyaging through strange seas of Thought, alone," as William Wordsworth imagined Isaac Newton doing. How, I wondered, would I even determine which direction to sail in, let alone whether something interesting awaited on the other side?

By 1999, I was two years into my PhD and painfully aware that I had yet to create any knowledge. That was when I started to notice a fresh new topic popping up repeatedly in talks and journal clubs and, perhaps most important, in the Cavendish's famous teatime conversations. Every day at 10:30 a.m. and again at 3:00 p.m., virtually everyone in the building would converge on the tearoom, where staff served tea and coffee starting at ten pence a cup. It was a tradition started by J. J. Thomson, who discovered the electron in 1897, and the semirandom seating meant that you'd find yourself chatting not just with your fellow students but also with grizzled lab technicians, visiting foreign academics, and eminent professors. The topic was usually either soccer or physics, and in the latter case I began to hear more and more about a mysterious entity called a quantum computer.

A computer, at its core, is a device that manipulates data stored in the form of "bits," each of which can take on a value of either zero or one. A quantum computer is one that obeys the peculiar laws of quantum mechanics. Most notably: instead of bits, it has qubits (pronounced "kew bits") that can exist in a superposition of zero and one simultaneously. Richard Feynman, a Nobel Prize–winning physicist, mused about the concept in a famous 1981 lecture at MIT. Four years later, the Oxford University physicist David Deutsch drew up a formal definition of how such a machine might work, and a few researchers around the world began to explore what it would be good for. In 1994, a mathematician at Bell Labs in New Jersey named Peter Shor published an algorithm showing that, if you could actually build a quantum computer, it would be able to solve certain problems like factoring large numbers—the core of modern encryption methods— exponentially faster than any classical computer.

It was Rolf Landauer, a physicist at IBM's research center in New York, who coined the term "Oxford flu" to describe the sudden surge of interest that followed Shor's breakthrough. What had once been an esoteric brain game played mostly by theoretical physicists and mathematicians affiliated with the University of Oxford became the talk of teatime in physics departments around the world. Grant money began to flow, including from organizations concerned with cryptography like the U.S. National Security Agency. My own research at that point was focused on the curious properties of certain exotic semiconductors, but I began to reconsider what these structures might be good for. I added a slide to my talks reframing my experiments as part of the grand pursuit of a quantum computer.

I still hadn't created any knowledge, though. I was following a trail blazed by Feynman and Deutsch and others. I didn't know how they had managed to find this potentially fruitful new territory to explore. Neither, it turned out, did Deutsch. "The stuff that I did in the late nineteen-seventies and early nineteen-eighties didn't use any innovation that hadn't been known in the thirties," he later told a reporter. "The question is why."

Coming up with new ideas is a more abstract concept than discovering a new continent, but the underlying processes of physical and conceptual exploration have more in common than you might think. "There's a lot of evidence that the same neural machinery we use for exploring the physical world around us is also leveraged to explore more abstract concepts," says Charley Wu, the computational cognitive scientist we met in chapter 6.

Some of the parallels are obvious: the explore-exploit dilemma, for example, recurs across domains. Just as European spice merchants once had to choose between navigating the well-known but arduous Silk Road or sailing west to seek a new route to Asia, companies have to choose whether to devote their research dollars to incremental improvements of their existing offerings or blue-sky attempts to come up with completely new products. Artists have to decide whether to create new work within the parameters of existing genres, or break the rules. Movie studios have to weigh the benefits of a fresh but untried story versus yet another sequel. Like pulling the lever in a multi-armed bandit experiment, choosing the uncertain option gives you a chance of dramatic success but also raises the likelihood of abject failure.

But the parallels run deeper than risk-reward calculations. Since the discovery of cognitive maps in our brains in the 1970s, scientists have been debating whether this neural circuitry is merely a sort of internal GPS system, or whether it has broader uses. The latter view—the intellectual offspring of Edward Tolman's idea that broad conceptual cognitive maps help us navigate "that great God-given maze which is our human world"—has gained the upper hand in recent years. We don't just map *places* in our hippocampus; we also map *ideas*. We keep track of our social networks, for example, by charting how near or far people are from us in a two-dimensional space defined by how powerful the other person is and what sort of experiences we have with them. That map too is plotted in the hippocampus.

One telltale sign of how we map ideas is the language we use to describe them. Spatial metaphors "structure some of our most fundamental concepts, including our concepts of time, quantity, similarity, good, and evil," psychologists Benjamin Pitt and Daniel Casasanto point out in a 2022 paper. "For example, in English, we use vertical space to talk about high and low numbers, lateral space to talk about the left and right poles of the political spectrum, and sagittal space to talk about moving meetings forward or back in time. Quantities can be big or small; vacations can be long or short; acquaintances can be close or distant." Even the internet is encoded as a physical space in our brains: we point our thumbs upward to signify approval, and click a leftward-pointing arrow to go "back" to a previous page.

The spatial organization of ideas isn't just a quirk of language. Similar patterns show up in unexpected and unspoken ways. For example, we tend to have more positive associations with words that contain more letters typed with the right hand than the left hand—an effect that has been demonstrated not just in English but also in Dutch, Spanish, Portuguese, and German, and occurs in both right-handers and left-handers. Clearly there's no evolutionary benefit to this "QWERTY effect." It's just a byproduct of how we map concepts like "good" and "bad" in our minds: if something is getting better, we tend to imagine it moving to the right on some imaginary scale. Ten out of ten is on the right (or at the top), while zero out of ten is on the left (or at the bottom).

We can even find shortcuts between ideas. A 1996 study by brain scientist Howard Eichenbaum trained rats to associate certain odors with other odors, or with rewards buried under a mix of sand and ground-up rat chow. He found that if rats learned to associate odor A with odor B, and also learned to associate odor B with the presence of a buried reward, then they could make the cognitive leap to assume that odor A also implied a reward. This is exactly analogous to the role of a cognitive map in physical wayfinding: if you know how to get from home to school, and from school to the library, your hippocampus enables you to plot a route directly from home to the library. Tellingly, Eichenbaum found that rats with a damaged hippocampus couldn't figure out the conceptual shortcut.

A 2020 study from Charley Wu and his colleagues tested the parallels between exploring in space and exploring ideas by having subjects complete two different treasure hunts. In the spatial treasure hunt, subjects used the arrow keys to navigate around a two-dimensional map searching for pockets of high reward, choosing whether to exploit whatever rewards they'd found in one region or explore other regions in search of richer yields. The conceptual treasure hunt involved using the arrow keys to change the shape and orientation of a striped geometrical pattern called a Gabor patch. Subjects had to explore the various possible combinations of features to figure out which ones had been assigned the highest reward value. In both cases, the subjects used a combination of random and uncertainty-directed exploration to zero in on the treasure. We search for new ideas in much the same way as we wander through the streets of an unfamiliar city, integrating clues from what we've seen before to predict what we'll find around the next corner, and tracking it all in our hippocampus.

Quantum computing's origin story is a perfect illustration of a common source of new ideas: the intersection of two old ones. The familiar idea, for David Deutsch, was quantum mechanics. Deutsch is a physicist's physicist, a famously eccentric Englishman who, despite his affiliation with Oxford, has never held an actual academic job there—largely because he hates teaching, but also because he prefers not to leave his cluttered but comfortable house. When a Japanese film crew once wanted to clean up the mess before filming an interview, they had to promise to take extensive photographs and reconstruct the chaos after they'd finished. His PhD, which he completed in 1978, dealt with quantum field theory in curved space-time. But it was a brush with ideas from a completely separate field—computer science—that pointed him in a new direction.

At the time, computer scientists were excited about the emerging new subfield of computational complexity, which seeks to understand and classify the difficulty of computational problems. Deutsch was skeptical of the whole endeavor. How could you measure the difficulty of a problem, he asked a colleague, without a universal standard computer to run the calculations on? "Well, the thing is, there *is* a fundamental computer," the colleague replied. "The fundamental computer is physics itself." Deutsch was struck by this insight—but it also occurred to him that computer scientists were still using the wrong computer. They were making their calculations based on classical Newtonian physics, which is a simplified approximation of the more complex rules of quantum mechanics fleshed out by physicists like Erwin Schrödinger and Werner Heisenberg in the 1930s.

Deutsch began to wonder what a universal computer—a notional construct devised by Alan Turing in the 1930s—would look like if it followed the rules of quantum mechanics. That question ultimately led to the 1985 paper in which Deutsch introduced the concept of a qubit and effectively created the field of quantum computing. Half a century had passed since Turing, Schrödinger, and Heisenberg had presented their respective ideas, but no one had connected them before. Similarly, Richard Feynman's early insights about quantum computing were prompted in part by his son, a student at MIT, switching focus from philosophy to computer science. Feynman already had the physics; now he was curious about computers, too. As with Deutsch, it was the relationship between two distinct bodies of knowledge—a newly charted pathway in his cognitive map of ideas—that generated Feynman's fresh insights. This is no coincidence, it turns out.

In 2017, a research team from Belgium and the United States published an analysis of all 785,000 articles that appeared in a widely used scientific database in 2001. They were interested in whether studies that connect previously separate bodies of knowledge are more likely to produce breakthroughs. They assessed "combinatorial novelty" by looking at whether a paper cited previous research from two or more journals that had never before been cited together in a single paper. The vast majority of studies in the database were exploitative rather than exploratory: just 11 percent made at least one new combination, and these novel studies initially got a cold reception. They tended to be published in lower-impact journals, and were less likely to be cited by subsequent papers in the first few years after publication. But after about three years, they caught up. Ultimately, the most novel one percent of papers were 40 percent more likely to end up as a "big hit," racking up large numbers of citations and influencing multiple fields. There are similar findings in the patent literature: filings that combine fields of expertise are more likely to produce breakthroughs. On the other hand, the novel papers were also more likely to end up as duds, rarely cited by subsequent researchers. Generating genuinely new ideas is a high-risk, high-reward enterprise.

In a sense, bridging the gap between two distant fields is the conceptual equivalent of looking for a new route between Europe and India, or crossing the uncharted interior of a continent. The start and end points are already known, but mapping the route between them creates new knowledge. Henri Poincaré, the brilliant nineteenth-century French polymath who is sometimes said to be the last person to know all of mathematics, argued that the greatest insights come from linking "domains which are far apart." That's easier said than done, of course. Big-data analyses of how we discover new music in online catalogs and how new pages get added to Wikipedia fit with a model of "expanding the adjacent possible." We tend to discover things on the border of what we already know, the researchers write, "mostly retracing well-worn paths, but every so often stepping somewhere new, and in the process, breaking through to a new piece of the space." We're like John McDouall Stuart, crossing Australia one waterhole at a time and never venturing too far from our last camp.

The other approach—call it the Burke and Wills strategy—involves greater leaps and greater risks. But the payoff is sometimes worthwhile. In 2015, a University of Chicago team analyzed millions of scientific articles and patents over a thirty-five-year period in order to map the network of relationships between pairs of chemicals and quantify the "cognitive distance" between them. The overwhelming majority of biomedical scientists, they found, stuck to "exploring the local neighborhood": if they made a new connection between two chemicals at all, it was only a slight variation on previous combinations. The biggest breakthroughs—those that won Nobel and other prizes and racked up the most citations—tended to come from bridging bigger cognitive distances, but relatively few scientists followed this approach. "When the concepts under study are more distant, more effort is required to imagine and coordinate their combinations," the researchers conclude. "More risk is involved in testing distant claims, because no similar claims have been successful."

The art of discovery, then, involves a mix of bold leaps and incremental progress: Burke and Wills *and* John McDouall Stuart. It was on a visit to the Van Gogh Museum in Amsterdam that Dashun Wang had an epiphany about how best to combine them. A former physicist who now heads Northwestern University's Center for Science of Science and Innovation, Wang is an expert on "hot streaks": sustained periods of higher-than-normal creativity or success. In a 2018 study, he and his colleagues analyzed the career trajectories of thirty thousand scientists, artists, and film directors, quantifying the impact of each individual work by the number of citations a scientific paper accumulated, a painting's most recent auction price, or a movie's IMDB rating. The key finding was that the best works in an individual's career tended to cluster together in relatively short streaks of a few years.

This finding was, in itself, surprising. Conventional wisdom says that people tend to do their best work in their thirties and forties, at the intersection of youthful enthusiasm and hard-earned experience. But Wang's data found no such pattern. Instead, career peaks were scattered more or less at random throughout the lifespan. Hot streaks didn't coincide with periods of high productivity, either. People didn't do *more* work during hot streaks; they just did *better* work. About 90 percent of the people in the database had at least one hot streak during their career, suggesting that it's a near-universal phenomenon. But nothing in the data seemed to indicate why they happen or what sparks them.

During Wang's visit to the Van Gogh Museum, he was struck by a shift in Van Gogh's style following his move to the south of France in 1888. That was a golden period in the painter's career—*The Yellow House, Starry Night over the Rhône, Van Gogh's Chair*—and Wang noted that it also seemed to coincide with a narrowing of painting style and subject choices. "If you look at his production before 1888, it was all over the place," Wang said. "It was full of still-life paintings, pencil drawings, and portraits that are much different in character from the work he created during his hot streak." Could this transition be a clue to the genesis of hot streaks, he wondered? He decided to crunch the data.

Quantifying the degree of exploration or exploitation in an artist's or scientist's work is a slippery task, especially when you're trying to analyze more than a million works from tens of thousands of individuals. Wang and his colleagues developed separate deep neural networks to analyze paintings, films, and scientific papers. Paintings, for example, were fed into an AI image recognition system that parsed low-level features like brushstrokes and line angles as well as higher-level features like shapes, objects, and art styles. This enabled each artwork's position to be plotted in an abstract two-hundred-dimensional space. An artist who is exploring will produce works that range widely through this 200D space; one who is exploiting will generate works that cluster together.

Sure enough, a pattern showed up in the data: hot streaks were most likely to occur following a sustained period of exploration. This was true for artists, scientists, and film directors alike. Jackson Pollock, for instance, was a struggling young artist in New York when the heiress Peggy Guggenheim offered him a monthly stipend of $150 in the early 1940s, freeing him to experiment with different approaches. That period of wide-ranging exploration persisted until about 1946, when he stumbled onto the drip-painting approach that he's now most famous for. Like Van Gogh, his focus then narrowed—or, in the terminology of Wang's AI system, the entropy of his work decreased dramatically—until about 1950, during which time he produced the work that remains most highly valued by the art market.

Wang's 2021 *Nature Communications* paper also traces the careers of film director Peter Jackson and chemistry researcher John Fenn. Jackson's early work bounced between genres like biography and horror-comedy before hitting his stride with the *Lord of the Rings* films. Fenn, who received the 2002 Nobel Prize for developing a technique for weighing molecules called electrospray ionization, is notable because of how late in his career his hot streak emerged. After decades of eclectic research on a broad range of chemistry topics at Yale University, he began developing electrospray ionization in the mid-1980s, shortly before the university's mandatory retirement policy curtailed his lab space when he turned seventy. The timing proved to be highly contentious, because Fenn maintained that the crucial work was done *after* Yale forced him out, making them ineligible to share in the patent and royalties. Fenn eventually lost the resulting court case in 2005, but his success is a reminder that hot streaks aren't just for the young.

What Pollock, Jackson, and Fenn—and, on average, the 2,128 artists, 4,337 directors, and 20,040 scientists in Wang's database—shared was a propensity to explore widely in their chosen vocations, followed by a dramatic narrowing of focus when they found a rich seam to mine. That timing is crucial. Exploring, on its own, doesn't accomplish much if you just keep wandering around indefinitely. Exploiting too has its limits. Ten thousand hours of single-minded focus and deliberate practice is only productive if you've found the right thing to focus on. Instead, it's the combination that is powerful. In Wang's database, the probability of starting a career hot streak at any given moment was lower than average following sustained periods of pure exploration or pure exploitation. Exploiting *then* exploring also lowered the chances of a hot streak. But exploring *then* exploiting—that was the magic combination, raising the chances of a hot streak by 15 to 20 percent. "If you just do one or the other, you don't get the full impact," says Jillian Chown, one of Wang's coauthors at Northwestern. "It has to be the combination of exploration followed by exploitation: experimenting in different areas, learning different domains and approaches, then really hunkering down and developing that body of high-impact work."

To me, Dashun Wang's data offers the clearest signal we have that exploration—intellectual foraging that ranges across a broad landscape of ideas and possibilities—pays off in tangible ways. So how do we put that insight into practice?

The easy answer is the Toucan Sam approach: Follow your nose! It always knows! That nagging feeling that you'd like to know more about something? The urge to find out what lies beyond the end of the trail? These are signs that exploration will be fruitful. That's the implicit message in Marc Malmdorf Andersen's slope-chasing theory of play. "One of the things that really characterizes play, possibly above all else, is that it's so fun and engaging and internally motivating," he says. You don't have to force kids to play: they follow their instincts, and in doing so they make choices that maximize their learning about the world.

That's a theme that recurs in other grand theories of conceptual exploration. Jürgen Schmidhuber, a giant in the field of artificial intelligence, has a wide-ranging "Formal Theory of Creativity, Fun, and Intrinsic Motivation" that defines fun as, in essence, a signal of how quickly you're improving your model of the world. "The growing infant quickly gets bored by things it already understands well, but also by those it does not understand at all," he writes, and so is "always searching for new effects exhibiting some yet unexplained but *easily learnable* regularity." What feels like fun guides the infant's learning—and, Schmidhuber argues, should guide the learning of AI systems. Similar arguments, he adds, explain why some works of art are viewed as more interesting or aesthetically pleasing than others. Contrary to some proposed theories of aesthetics, it's not an objective property of the artwork. Instead, it's a function of the balance between predictability and surprise in the eye of the beholder—which explains, for example, how Stravinsky's *Rite of Spring* went from outré to orthodox.

But there's a problem. James March, the management theorist whose much-cited 1991 paper galvanized interest in the explore-exploit trade-off, argued that we don't explore as much as we should, at the level of individuals, organizations, and societies. The returns on exploitation, March writes, are "positive, proximate, and predictable," in contrast to exploration, whose returns are "uncertain, distant, and often negative." It takes three years, we saw earlier, before highly novel scientific research catches up on citation count, and it produces more duds along the way. As a result, the abstract feedback urging us to explore—the faint voice that says "Wouldn't it be interesting to try that?"—is drowned out by the clear and actionable feedback urging us to exploit. CEOs have only a few years to prove their worth; politicians and policymakers have even less. How can they resist? The result is a Rousseauvian conflict between nature and civilization: humans are born to explore but are everywhere chained by the demands of the next annual report, the next election, or the need to avoid inconvenient gaps on the CV.

By the time I finished my PhD, I hadn't discovered any new continents of knowledge, but I'd mapped a few backstreets in an obscure subdivision. It was enough to get my degree. The Oxford flu, meanwhile, had morphed into a full-fledged global pandemic. After my brief attempt to make it as a professional runner fizzled, I leaned into my professional rebranding and in January 2002 started a postdoctoral position at a National Security Agency lab affiliated with the University of Maryland, in their new quantum computing research group.

The NSA's interest in quantum computing was twofold. First and most obvious, they wanted to "read Osama's email," as one of my bosses put it when I was interviewing for the job. If they could build a quantum computer capable of implementing Shor's algorithm, they would be able to hack into a large fraction of the world's encrypted communications. This was a blue-sky goal that most scientists at the time figured wouldn't be possible for decades, if ever. But the NSA also had a defensive mandate to protect the privacy of U.S. government communication, with a fifty-year time horizon. They needed to ensure that no one else could read George W. Bush's encrypted cables even in 2052. To that end, they needed to know well in advance whether a working quantum computer would *ever* be feasible.

The biggest barrier to building a working quantum computer is called decoherence, which is the transition from quantum to classical behavior. The laws of quantum mechanics tell us that an electron can be in two places at once, and can do other unexpected things like teleporting from one location to another. The laws of everyday experience tell us that a baseball can do none of these things. Building a working quantum computer would require assembling a system the size of a baseball that nonetheless behaves in some ways like an electron. But no one was—or is—really sure what defines the boundary between baseballs and electrons. Is there a fundamental law of nature that forces objects to decohere—to collapse from their delicate quantum states into the more familiar classical modes—beyond a certain size or complexity?

My training as a semiconductor physicist suited me for the practical job of trying to build a quantum computer. But I found the basic physics questions much more intriguing, so I followed my nose. A senior NSA researcher named Keith Schwab was exploring these questions experimentally by trying to see how big a structure he could build that would obey the electron-like properties of quantum mechanics. I began to work with Keith's group on his "nanoelectromechanics" approach, which included carving tiny freestanding bridges just a few hundred nanometers wide—a thousand times narrower than a human hair—out of a silicon chip and then using an electric field to vibrate them back and forth. The goal was to put the bridge in a superposition of two quantum states, bending its physical structure in two opposite directions at the same time.

I mostly tried to ignore the growing disconnect between the abstract questions about quantum mechanics that we were exploring in the lab and the practical code-breaking machine that the NSA's budget makers thought they were buying with their tens of millions of dollars of funding. We couldn't necessarily trace a direct path from vibrating silicon nanobridges to a future quantum computer, but we were confident that acquiring a deeper understanding of quantum mechanics would prove to be worth the effort. Our attitude echoed that of Michael Faraday, who made crucial discoveries about the basic properties of electromagnetism starting in the 1820s. When William Gladstone, later the prime minister of Britain, asked Faraday what his esoteric research could possibly be useful for, he replied, "Why sir, there is every probability that you will soon be able to tax it!"

That tension between basic and applied research is another example of James March's tug-of-war between exploration and exploration—and it's not a new phenomenon. In 1937, for example, the educational reformer Abraham Flexner delivered a now famous lecture at Bryn Mawr College (later published as an essay in *Harper's Magazine*) titled "The Usefulness of Useless Knowledge," in which he argued that "throughout the whole history of science most of the really great discoveries which had ultimately proved to be beneficial to mankind had been made by men and women who were driven not by the desire to be useful but merely the desire to satisfy their curiosity." Flexner was the founding director and key architect of the Institute for Advanced Study, a highly unconventional research center in Princeton, New Jersey, endowed by grocery-store magnates Louis Bamberger and Caroline Fuld in 1930. He lured world-famous thinkers including Albert Einstein, John von Neumann, and Kurt Gödel with the promise of complete freedom from any responsibilities or expectations other than those suggested by their own genius. "No faculty meetings are held," he boasted. "No committees exist."

Flexner cited the example of Paul Ehrlich, who in the 1870s was a mediocre medical student at the University of Strasbourg. His anatomy professor, Wilhelm von Waldeyer, later recalled that the seventeen-year-old Ehrlich showed little interest in the dissections that he was supposed to be learning, and instead spent his time peering through his microscope and covering his desk with colored dots:

> As I saw him sitting at work one day, I went up to him and asked what he was doing with all his rainbow array of colors on his table. Thereupon this young student in his first semester supposedly pursuing the regular course in anatomy looked up at me and blandly remarked, "*Ich probiere.*" This might be freely translated, "I am trying" or "I am just fooling." I replied to him, "Very well. Go on with your fooling."

Ehrlich squeaked through his medical training, mainly because it was obvious to his professors that he had no interest in treating patients. But he went on to a distinguished career in research, discovering the first effective treatment for syphilis and earning the 1907 Nobel Prize for his work on the immune system. And the colored dots on his desk? The forerunner of his techniques for staining bacteria and blood cells to identify them, which have had a transformative impact on modern medicine.

"*Ich probiere*" might well be the motto of the Institute for Advanced Study, and over the years its members have included thirty-five Nobel Laureates and forty-four of the sixty-two winners of the Fields Medal in mathematics. But even in Flexner's time there was an undercurrent of pressure to steer its faculty's research toward solving practical problems. "I dissent from your fundamental proposition that if a group of scholars could be comfortably housed, something of significance to thought and mankind will happen," the historian Charles Beard warned Flexner. "Death, intellectual death may happen, as it did in many a well-appointed monastery in the Middle Ages." Exploration, Beard felt, has to be balanced with exploitation.

In the years since the Institute's founding, the currents have continued to shift. Scientists played a crucial role in the outcome of World War II, not just through the Manhattan Project but also through practical innovations with lasting peacetime impacts like radar and jet engines. As Cold War rivalries simmered, the space race heated up, and the computer age began, the focus on science's tangible deliverables intensified. Paul Forman, a historian of science at the Smithsonian Institution, argues that by around 1980 a transition had occurred, corresponding to the shift from modernity to postmodernity: instead of technology being subordinate to science, the opposite was now true. Basic curiosity-driven exploratory research, from society's point of view, is now less important than the pursuit of new gadgets.

The portion of the U.S. federal budget devoted to research and development is currently around 0.8 percent of gross domestic product, as compared to nearly 2 percent in 1964. And even those figures don't break down the diminishing share of research dollars allocated to blue-sky research of the type that Michael Faraday or Paul Ehrlich might recognize. "No one wants to fund wild new ideas," the chief medical officer of the American Cancer Society admitted to the *New York Times*. This is the trend, writ large, that James March warned about. By some estimates, every dollar spent on basic research returns as much as eight dollars in downstream benefits. It's the best investment money can buy; but because the benefits are "uncertain, distant, and often negative," it's the easiest thing to cut when budgets are tight—as they always are.

In October 2019, Google announced that its quantum computing research team had achieved a long-sought milestone: quantum supremacy. Their test machine, consisting of just fifty-three qubits made from loops of superconducting wire cooled to a hundredth of a degree above absolute zero, took two hundred seconds to perform a calculation that they estimated would take ten thousand years for the world's most powerful supercomputer. A rival team from IBM immediately disputed the claim, claiming that a supercomputer could actually get the answer in just 2.5 days. But the exact numbers don't really matter. For a certain subset of carefully chosen test problems, quantum computers had finally surpassed conventional ones, and in the years since 2019, Google and several other companies, including upstart rivals in

China and Canada, have announced ever-stronger claims of quantum supremacy.

The field has come a long way since I left my NSA postdoc in 2004 to head to journalism school. At that point, the benchmark of success was a 2001 experiment that used a seven-qubit quantum computer to determine—with moderate certainty—that the factors of fifteen are three and five. That's not to say that quantum computers will be on the shelves anytime soon. *Nature* put Google's 2019 announcement on the cover of its 150th anniversary issue, but the journal included an enthusiasm-dampening editorial warning that everyday quantum computers, which could radically accelerate tasks like drug discovery, climate modeling, and complex logistical planning, were still decades away at best. The technology, the editorial warned, is "still near the start of a long and unpredictable journey." That remains true as I write this.

But there's one other thing that such journeys have in common with the journeys of explorers in the physical world: sometimes knowing where you're headed isn't an advantage. Alexander Mackenzie never would have set out for the Pacific Coast if he'd realized how far away it actually was; neither would Columbus have set out for India. The same can be true in the world of ideas. "I never would have conceived my theory, let alone have made a great effort to verify it, if I had been more familiar with major developments in physics that were taking place," the great Hungarian-born physical chemist Michael Polanyi wrote about one of his scientific contributions. That's true in life more generally, whether you're brainstorming ideas at work or, like Nils van der Poel, rethinking how to train for your next race. It's easy to find arguments for sticking with the familiar; but sometimes you have to venture into the unknown and see for yourself what's out there.

Part III

WHAT EXPLORING MEANS NOW

9

THE PROBLEM WITH PASSIVE

Nellie Bly, the *New York World*'s star investigative reporter in the 1880s, had just one story idea to present at her weekly Monday-morning pitch meeting.

"I want to go around the world," she said, settling into her chair.

"Well?" replied her editor, Jack Cockerill.

"I want to go around in eighty days or less. I think I can beat Phileas Fogg's record. May I try?"

Cockerill was an imposing figure. The owner of the *World*, Joseph Pulitzer, had installed him as editor in chief after he'd shot and killed an angry reader in his previous position at the *St. Louis Post-Dispatch*, which Pulitzer also owned. But he responded positively to Bly's proposal. Jules Verne's 1873 novel *Around the World in Eighty Days* had been an international bestseller, so Fogg's fictional journey was familiar to readers. And it wasn't entirely implausible: the opening of the Suez Canal in 1869 had created a direct passage for ships to travel between Europe and Asia, radically shortening the time needed for a round-the-world trip. Indeed, a rich businessman named George Francis Train had done it in 1869 and 1870, spending exactly eighty days traveling—though his total time away was longer because he spent two weeks in jail in France after supporting a short-lived revolution there.

There was a problem, though: Bly was a woman. In fact, the paper's editors had already considered the possibility of sending a reporter around the world, and had concluded that it was a man's job. Cockerill knew Bly's abilities, so he suggested they discuss it with George

Turner, the *World*'s business manager. Turner, however, was emphatic. "It is impossible for you to do it," he said. "In the first place you are a woman and would need a protector, and even if it were possible for you to travel alone you would need to carry so much baggage that it would detain you in making rapid changes."

Bly was understandably indignant. Her very first assignment for the *World* had involved feigning insanity to get herself locked inside the Blackwell's Island Insane Asylum for ten days in order to expose the mistreatment of female patients there. She had subsequently gone undercover in sweatshops and chorus lines, allowed herself to be lured into the carriage of a rumored predator in Central Park to expose him, arranged to sell a (fictitious) baby, and purchased the committee votes of corrupt state legislators. She was comfortable traveling alone.

"There is no use talking about it," Turner said, trying to close the conversation. "No one but a man can do it."

"Very well," replied Bly. "Start the man, and I'll start the same day for some other newspaper and beat him."

Why did Nellie Bly want to travel around the world? The obvious answer is that she wanted to find out whether it was possible to make it in eighty days. If that was her main motivation, then it wouldn't have mattered if the *World* sent a male reporter instead. Either way, she would get an answer. She presumably wanted to see what other countries were like; but in that case too she could have read the dispatches the lucky reporter sent back from his trip. It's also reasonable to assume that she wanted to make her name and advance her career. But in reading her account and assessing the overall arc of her career, you get the sense that she really wanted to make the trip for its own sake. And who can blame her?

We can push the thought experiment a little further, along the lines of Harvard philosopher Robert Nozick's famous "experience machine." Imagine a virtual reality machine capable of simulating any experience you desire. "Superduper neuropsychologists could stimulate your brain so that you would think and feel you were writing a great novel, or making a friend, or reading an interesting book," Nozick wrote. "All the time you would be floating in a tank, with electrodes attached to your brain." In other words, you're in the Matrix. Nozick proposed the scenario in 1974 to counter the idea that we value pleasure above all else, under the assumption that most people would prefer to live in the imperfect real world than to stay plugged in for the rest of their lives. Would Bly have been satisfied with an experience machine simulation of a trip around the world?

Predictive processing says that we're driven to explore in order to resolve uncertainty and acquire information. In the experience machine, Bly would resolve all the same uncertainty and acquire all the same information as the person who took the actual trip. But that still doesn't seem like an adequate substitute. There's a difference between actively embarking on your own adventure and passively tagging along on someone else's; between sitting in the driver's seat and staring out the window of the passenger seat; between being a protagonist and a spectator. Our brains process and respond to these two types of experiences differently, even when the information we encounter is identical. And this is a problem, because a variety of societal and technological forces are pushing us toward passive rather active exploration, with repercussions for our happiness, our productivity, and perhaps even our health.

This shift from active to passive is just one facet of a much larger societal transformation. In his 2000 book *Bowling Alone*, Harvard social scientist Robert Putnam charted the decline of community involvement from the 1960s onward in the United States. Putnam was primarily worried about social cohesion and the future of democracy as bowling leagues and Boy Scout troops and Rotary Clubs dwindled. These days, even people who are passionate about social or political or environmental causes tend to be involved only passively. They may join organizations like the National Wildlife Federation or the American Association of Retired Persons, but they're not marching in the streets or even getting together for monthly meetings with the local chapter. "For the vast majority of their members," Putnam pointed out, "the only act of membership consists of writing a check for dues or perhaps occasionally reading a newsletter."

Putnam also charted the decline of town bands and amateur dramatics and community sports teams. A century ago, your Friday night entertainment was provided by your neighbors; these days, instead of modestly talented amateurs, you can listen to the greatest musicians in the world and watch the greatest athletes on earth. With that competition, who needs the town band? It's not that people are less interested in music or drama or sports—on the contrary, *consumption* of various forms of both popular and high culture continues to grow. But participation—undertaking the act of creation, accepting the risk of competition—is drifting downward. That's a problem, Putnam wrote: "In football, as in politics, watching a team play is not the same thing as playing on a team."

Mihaly Csikszentmihalyi, the psychologist who coined the idea of flow as an immersive and enjoyable state of focus, voiced similar concerns: "Instead of using our physical and mental resources to experience flow, most of us spend many hours each week watching celebrated athletes playing in enormous stadiums. Instead of making music, we listen to platinum records cut by millionaire musicians. Instead of making art, we go to admire paintings that brought in the highest bids at the latest auction. We do not run risks acting on our beliefs, but occupy hours each day watching actors who pretend to have adventures, engaged in mock-meaningful action."

It's that last point that connects Putnam's concerns to ours. To explore the world, you have to be *in* the world, capable of acting on it and being acted upon by it, making decisions whose outcomes are unknown and have consequences. Taking a penalty kick or playing a concerto are miniadventures whose endings aren't yet written. Csikszentmihalyi's distinction reminds me of all the times I wandered through arcades as a kid with no quarters in my pocket. I'd stop to "play" some of the games, moving the joystick and punching the buttons as the game's demo mode cycled through its various worlds and showed me the adventures I could be having. But the joystick and buttons didn't do anything; I wasn't causally connected to these worlds, so it got boring very quickly. "The flow experience that results from the use of skills leads to growth," Csikszentmihalyi wrote; "passive entertainment leads nowhere."

Nellie Bly, for one, was determined to go somewhere. She was born in a small Pennsylvania coal-mining town in 1864. Her father died when she was six, leaving her family of six nearly destitute; her mother then married an abusive drunk and divorced him five years later. Bly had to drop out of school at fifteen to help support the family working as a kitchen girl and taking a succession of other menial jobs. Five years later, she wrote an angry letter to the editor of the *Pittsburgh Dispatch* about a column that criticized modern women who dared to work outside the home. The editor didn't print the letter, but he did one better: he hired her as a reporter at a salary of five dollars a week. He wanted her to write about "women's issues" like shopping and fashion; instead, she wrote about working conditions for women in factories and convinced the editor to send her to Mexico as the paper's correspondent there. When she returned five months later, she couldn't face being banished back to the women's page. One day she simply didn't show up for work. A colleague found a note on his desk: "I am off for New York," it read. "Look out for me. BLY."

When she initially broached the possibility of a round-the-world trip in 1888, Bly was twenty-four and had been in New York for three years. At first nothing came of it, but a year later, on a cold, wet November evening, she was suddenly summoned to her editor's office. "Can you start around the world day after tomorrow?" he asked. "I can start this minute," she replied. The next day was filled with frantic preparations. She went to a dressmaker to have a durable traveling dress made; she bought a single handbag as luggage, determined to disprove the idea that women couldn't travel light; she had an emergency passport drawn up by the secretary of state himself and couriered from

Washington to New York by a *World* correspondent. And she paid a visit to the Broadway offices of the international travel agency Thomas Cook & Son, to plot a round-the-world itinerary.

If there's one person who can be credited with ushering in the era of modern mass tourism—of bridging the gap between medieval pilgrims and Chevy Chase in *European Vacation*—it's Thomas Cook, a nineteenth-century businessman and Baptist preacher in the English Midlands. In 1841, Cook organized a special train to take 485 people from Leicester to nearby Loughborough for a temperance meeting, at a cost of one shilling a head. The trip was a success, and Cook continued to organize excursions over the next few years, harnessing the relatively new railway system in service of the temperance cause. In 1845, he decided to turn this sideline into a business, organizing trips farther afield to places like Scotland and Ireland. During the summer of 1851, he organized the transport by train of some 150,000 people from northern England to and from London's Great Exhibition.

In the years that followed, Cook began organizing tours to Europe, North America, and Africa. Eventually, in 1872, he led his first around-the-world tour: it was seeing a newspaper advertisement for this trip that sparked Jules Verne's idea for *Around the World in Eighty Days,* though Cook's eleven-person tour group ended up taking 222 days. Among Cook's innovations were the idea of package tours, in which a fixed price would cover transport, accommodation, and food (the round-the-world itinerary was billed as costing 270 guineas, not including wine, beer, and spirits); traveler's checks that could be redeemed for local currency at the destination; and concise guidebooks filled with practical local advice. The concept of a "Cook's tour" entered the language (it's still in the *Oxford* dictionary), and the company remained the preeminent force in mass tourism well into the twentieth century.

Whether Cook made travel more active or more passive isn't a straightforward question to answer. "At that time, many people had never ventured farther than a few miles from their birthplace," the historian Stephen Usherwood points out. "This was due as much to lack of curiosity and self-confidence as to want of money." Usherwood's claim about a lack of curiosity is debatable, but it's undoubtedly true that very few people had the know-how to undertake international travel at the time. Breaking down those barriers is perhaps Cook's greatest accomplishment and remains the primary goal of the vast infrastructure of modern tourism. You can think of it in terms of the Wundt Curve, in which intermediate levels of uncertainty are the most satisfying: Cook helped move vast swaths of the globe from the far side of the uncertainty curve into the sweet spot.

On the other hand, he also created the template for an increasingly passive form of travel, in which you can be ushered through even the most remote landscapes without making any decisions, without engaging or interacting with local people, without running any risks. The pinnacle of this sort of passively active exploration, both literally and figuratively, is the modern Mount Everest industry. "You still have to climb the mountain with your feet," acknowledged Tenzing Norgay's son Jamling Tenzing in 2003, fifty years after his father's 1953 summit expedition. "But the spirit of adventure is not there anymore. It is lost. There are people going up there who have no idea how to put on crampons. They are climbing because they have paid someone $65,000."

I've always preferred to find my own way rather than hire a guide, even in places like Nepal, where guides are inexpensive and strongly recommended. (That may no longer be an option, as Nepal banned trekking without a guide in 2023.) The only exception for me was on a 2010 trip to Papua New Guinea, where my wife and I hired a pair of guides to lead us along the Kokoda Track, through the remote Owen Stanley Mountain Range that divides the north and south coasts of the country. The route passes through privately owned tribal land, so trekking unaccompanied is a serious cultural faux pas. We enjoyed our guides' company and learned far more about local history and culture than we would have on our own—but the experience was very different from other backcountry trips we've done alone. We were passive passengers, rather than actively charting a course across the mountains. On the vast spectrum between watching a *National*

Geographic documentary on TV and exploring an unknown wilderness, having guides tilted the experience toward the former.

To be clear, there are times when having a guide makes sense, depending on where the adventure in question lies on your personal Wundt Curve. (There are also times when watching a *National Geographic* documentary is a fantastic way to spend an evening.) But not always, and not for everyone. Back in 1981, leisure studies scholars Leo McAvoy and Daniel Dustin published an article called "The Right to Risk in Wilderness" in the *Journal of Forestry*, proposing the establishment of "no-rescue" wilderness zones in remote places like Alaska's vast Gates of the Arctic, the least-visited national park in the country. "People are drawn into experiences they probably shouldn't be even doing because they haven't invested their own energy and time in developing the skills to take care of themselves," Dustin told me. "They outsource their expertise to a guide, or they outsource it to a cell phone, or a GPS, so that if something bad happens, they can be sure someone will come and bail them out."

No-rescue zones are the logical but extreme endpoint of my aversion to guides. Their existence would force those who chose to venture into them to weigh every decision with utmost care. In a small but significant way, they would strip away some of the illusion of adventure and replace it with the unvarnished real thing. The concept has sparked plenty of debate, but it has never gained official approval—not in public, at least. Dustin recalls being ridiculed for the proposal by the chief of the U.S. Forest Service during a conference of search and rescue professionals. Afterward "we found ourselves standing next to each other at the urinals, and he elbowed me and said, 'Dan, just between the two of us, I think it's a great idea.'"

Much of the debate, understandably, has focused on what could go wrong. Would we really stand by and let lost or weather-tossed hikers die? But to Dustin, who once walked sixteen miles back to a trailhead in Wyoming's Wind River Range on an ankle broken in three places, it's more interesting to think about what would go right for the vast majority of people. You head into the wilderness, you face unexpected challenges, and you discover that you have within you what's needed to surmount them. But that's only possible if the outcome isn't predetermined. "It's the uncertainty of all of it, it's the question marks, it's not knowing what's going to happen," he says. "It's not having a map that's filled in for you."

In reality, I suspect vanishingly few people would choose to visit a hypothetical no-rescue wilderness. I certainly wouldn't, and even the most intrepid polar explorers can and do call for emergency extraction when necessary these days. "Publicly you could never say this is a good idea, because it sounds so heartless and cruel," Dustin says. "My point is in our culture, we will glorify a few experts who do this sort of thing, and we'll put 'em on television, maybe create a reality TV show or whatever, but if everyman wants to do it, stretch herself or himself, we somehow suggest that that's just ridiculous." Wilderness adventure, like music and sports, has become a passive spectacle rather than an active pursuit for most of us.

Is the shift from active to passive, and from participating to spectating, really so bad? Aren't we better off listening to Yo-Yo Ma and Taylor Swift rather than the local middle school marching band, and listening to David Attenborough narrate stunning footage of the Arctic tundra rather than risking our lives dodging polar bears in a blizzard with no possibility of rescue? Sure, in most cases. But the real distinction between active and passive play isn't swashbuckling through the wilderness versus lounging on the sofa. It's about autonomy. Do you have the freedom to follow your nose, to shape your experience by choosing to pursue the paths—literal or intellectual—that interest you the most? Or are you just swallowing whatever you've been spoon-fed?

By that metric, the problem with social media isn't just that we spend so much time sedentary and staring at a screen—it's that we surrender too much autonomy to the algorithms that choose which content to serve us. These personalized feeds signal "the death of exploration, trial and error and discovery," according to Chris Murphy, a Connecticut senator who cosponsored a bipartisan 2023 bill to limit children's access to social media. "Why should students put in the effort to find a song or a poem they like when an algorithm will do it for them? Why take the risk to explore something new when their phones will just send them never-ending content related to the things that already interest them?"

Similar trends are also altering how we explore in the physical world. "The Google map on the phone in your pocket will show a kind of biased view of the world, tilted towards the things that it knows you are interested in based on your searches and also the businesses you've visited," says Colin Ellard, a University of Waterloo psychology professor who studies how we navigate and how the spaces around us affect our brains and bodies. In a sense, it's a physical manifestation of the idea of "filter bubbles," the worrying social media trend in which clever algorithms present us with only the news and information we already agree with. "To me," Ellard says, "something feels wrong about each of us inhabiting our own personal map rather than sharing something more universal."

There's research to back up the idea that being fed information, rather than having to seek it, stifles exploration. A 2011 study led by Elizabeth Bonawitz, who now leads Harvard's Computational Cognitive Development Lab, gave preschoolers a handmade toy cobbled together in the machine shop from colorful PVC pipes. The toy had various hidden features: pulling on one tube made it squeak, pressing on others made them light up or play music or reveal an upside-down mirror. When the experimenters taught one group of kids that it could squeak, they would play with it for a little while until they figured out how to make it squeak, then get bored. A different group of children weren't taught how the toy worked, but saw the experimenters make it squeak seemingly by accident. This latter group spent longer playing with the toy and were more likely to find its other features too.

Some of the differences here are a consequence of *how* the children chose to explore in the two scenarios, Bonawitz says. When an adult taught them about one of the toy's features, the kids inferred that this particular feature was all they needed to know about the toy. Similarly, if a guide is leading you up a mountain, you might reasonably assume that the path he's leading you on is the only path you need to pay attention to, and the points of interest he flags are the most important ones. That doesn't mean that teaching or guiding are bad things, Bonawitz says. Teaching is a great way of helping learners zero in on key ideas, and there are ways of providing open-ended guidance that stimulate exploration rather than shutting it down. But in this particular scenario, leaving the kids to figure out how they wanted to play with the toy led them to explore it more thoroughly.

That's only part of the story, though. "If you're exploring more, you'll discover more," Bonawitz says. "But even if both groups have the same information presented to them, either by someone else or by their own discovery, there's also evidence that being actively involved helps you encode it better." This apparent difference between active and passive learning has been a topic of long-standing debate among education theorists, but neuroscientists have also begun to weigh in. A famous experiment in the 1960s yoked a pair of kittens together on a circular merry-go-round, allowing one of them to propel itself freely while the other's motion was dictated by the movements of the first one. Even though they both saw the same things, the active kitten learned more effectively. Half a century later, a human version of the experiment asked subjects in a brain scanner to memorize a grid of objects through a narrow window whose position was controlled either actively or—like the kittens—by the actions of another subject. Active exploration was more effective, even though subjects saw exactly the same things in both conditions, and it triggered a distinct pattern of brain activity centered in the hippocampus.

Even in the context of the multi-armed bandit tasks used to study explore-exploit decisions, passive exploration yields fundamentally different—and inferior—results compared to active exploration. That's an insight that Robert Wilson stumbled upon accidentally when he moved from Princeton to take up a professorship at the University of Arizona in 2015. The Horizon Task that Wilson developed at Princeton starts with a series of forced choices: the screen tells you to pull the lever on the left, then the lever on the right, and so on, before allowing you to start making your own choices. This allows the researchers to control exactly how much information you have about each option.

But it's also time-consuming, so when Wilson began running experiments in Arizona he decided to simplify the protocol: instead of forcing subjects to play the machines themselves, he simply gave them the results of the first four pulls.

The problem with the new protocol was that it didn't work. Subjects no longer made rational choices about when to explore and when to exploit. They didn't seem to be learning from the information when it was passively fed to them, compared to when they played the machines themselves—even though the information was once again identical in the two situations. There were more subtle differences too: the subjects were more ambiguity-averse in the passive protocol, which means that they were less likely to explore in general. Wilson eventually realized what was happening, switched back to the active protocol, and started getting the expected results again.

Nellie Bly started her twenty-eight-thousand-mile voyage at 9:40 a.m. on November 14, 1889, when the steamship *Augusta Victoria* pushed off from the docks in Hoboken, New Jersey. If all went well on the Atlantic crossing, she would arrive in Europe with just enough time to pay a quick visit to Jules Verne at his home in Amiens before catching the weekly mail train to Brindisi, at the southwestern corner of Italy. The plan was tentative: no one seemed to be entirely sure whether the mail train actually ran every week, or whether the steamship it would connect with in Brindisi was headed for India or China. At 6:00 p.m. that evening, meanwhile, another young reporter named Elizabeth Bisland was boarding a train from New York's Grand Central Depot bound for Chicago. Her editor at a fledgling magazine called *The Cosmopolitan* had called her in only that morning; now she was racing Nellie Bly around the world, traveling in the opposite direction.

Bisland's itinerary too had been plotted a few hours earlier by Thomas Cook & Son. Over the weeks and months that followed, the two women raced from connection to connection, traveling mainly by railway and steamship. Bly's eastward route meant that the final leg of her journey would involve crossing the Rockies in the middle of winter, with the possibility of train tracks being blocked by snow. Bisland's westward route would finish with an equally daunting winter crossing of the North Atlantic. And there would be plenty of other opportunities for their plans to go awry. In Hong Kong, where Bisland was on pace for a round trip of less than seventy-three days, the record-setting steamer she planned to take to Ceylon (as Sri Lanka was then known) broke its screw on the way into the harbor, forcing her to switch to a slower ship. Bly too got hung up for five days in Ceylon when her connecting ship was delayed.

Five days in Ceylon might sound like an invitation to explore, but Bly didn't seem interested. In her book about the voyage, most of her account of Ceylon is devoted to retelling at great length the plot of a play she saw. She found the local temples uninteresting, and the nearby mountain town of Kandy underwhelming. She didn't visit the island's famous tea or cinnamon plantations. "In Ceylon," Matthew Goodman writes in his account of Bly's and Bisland's race, "Bly's renowned reportorial instincts deserted her." She was undoubtedly preoccupied by the race against time, and exhausted from the long weeks of travel. But you also get a sense that her enforced passivity was beginning to grate: for all the autonomy she had, she might as well have been stuffed in a parcel and mailed around the world.

Bly's trip was certainly easier than that of her predecessor Ferdinand Magellan, the Portuguese seafarer who, starting in 1519, led the first expedition to circumnavigate the world. By the time Magellan located the strait that now bears his name, which connects the Atlantic and Pacific oceans across the southern tip of South America, he had already lost one of his five ships in a storm and suppressed a bloody mutiny. When the strait split into two, he sent one ship up each branch, giving them four days to reconnoiter. One of the ships never returned and instead secretly sailed back to Spain. Magellan pressed on with his remaining ships and finally reached the Pacific Ocean, which he named for its seemingly calm waters. He figured it would take him three or four days to cross it; instead, he didn't reach land for another ninety-eight days, by which time twenty-nine sailors had died of scurvy and the survivors were eating charred leather and paying a half-*ducado* apiece for the rare luxury of a rat. The situation didn't improve from there: Magellan himself was killed in the Philippines while trying to forcibly convert some of its Indigenous residents to Christianity. Of the five ships and 270 men that set out from Spain, just one ship and eighteen men, under the command of Juan Sebastián Elcano, made it back in 1522.

The expedition is often hailed as one of the greatest sea voyages of all time, and it was certainly active rather than passive, but it was utterly horrific, even for those who survived. Given the choice, most of us would opt for Bly's ennui over Magellan's ordeal. That's an extreme version of the trade-offs we face whenever we weigh active versus passive forms of engagement. Cooking a meal from scratch is more rewarding than microwaving a frozen dinner but takes more effort. Setting up a board game to play with your kids is harder than chilling with Netflix: Robert Putnam figured that about a quarter of the decline in community engagement that he observed could be blamed on television. And when it comes to exploring, as I found when I hiked the Long Range Traverse in Newfoundland, it's far easier to let your phone's GPS guide you than to struggle with maps and compasses and confusing game trails.

I first met Véronique Bohbot in 2009, when I was writing a magazine feature about a woman with no sense of direction. The woman had been tagged as "Case one" of a newly identified neurological disorder called developmental topographical disorientation, and tests had shown that she couldn't form cognitive maps. When placed in a brain scanner and asked to navigate around a virtual neighborhood, the woman's hippocampus remained dormant, even though it worked perfectly during basic memory tests. Bohbot was one of the McGill University brain scientists, along with Giuseppe Iaria, who had published early studies on the difference between navigating with cognitive maps, using the hippocampus, and with stimulus-response cues, using the caudate nucleus. "Case one," for reasons no one understood, seemed to be all caudate nucleus.

At the time, the distinction between active and passive navigation was undergoing a dramatic change. GPS units were found mostly in cars, or in bulky handheld units that backpackers could use to check their coordinates. But just a year earlier, Apple had added GPS to its iPhones. There was already concern that outsourcing navigation to a gadget might somehow compromise our innate abilities and perhaps even change our brains. Bohbot had shown that people who default to cognitive map-based navigating in her tests tend to have more gray matter in their hippocampus, while those who default to stimulus response have more in their caudate nucleus. Moreover, the two regions seem to directly compete with each other: people with bigger hippocampi tend to have smaller caudate nuclei, and vice versa.

Developmental topographical disorientation struck me as an interesting neurological curiosity, like a case study in an Oliver Sacks book. But as I chatted with Bohbot, I got the impression that she saw it as a warning rather than a novelty. "Society is geared in many ways toward shrinking the hippocampus," she told me. The problem, in her view, was broader than GPS. She saw a societal shift toward instant gratification, efficiency at all costs, productivity as the only measure of value—all of which favor passive stimulus-response-driven thinking rather than slower, more effortful cognitive mapping. And a smaller hippocampus, she warned, raises your risk of various maladies, including Alzheimer's disease, depression, schizophrenia, and post-traumatic stress disorder.

It's one thing to concede that spending years memorizing every back alley and shortcut in London might produce subtle changes that show up in a taxi driver's brain scan; it's another to suggest that seemingly innocuous behaviors like following your phone's directions to a restaurant are going to give you dementia. But when I called Bohbot in 2024 to see how her views had evolved, she was even more worried. The trends she'd seen in 2009 have only intensified; and the evidence that these trends might be harmful has gotten stronger.

In one set of studies, Bohbot and her colleagues found that people who play a lot of video games tend to have a smaller hippocampus and default to stimulus-response navigation. But the type of video game matters. First-person shooters like *Call of Duty*, which often include GPS-like map overlays to tell you exactly where you are and where you need to go, tend to trigger the caudate nucleus. More exploratory games like *Super Mario 64* trigger the hippocampus: in one study, Bohbot found that playing *Super Mario 64* actually boosted hippocampus volume in older adults. And how you approach the game matters too. Bohbot had volunteers come into the lab to play various video games for a total of ninety hours over two to three months (while being paid nine dollars an hour!). Those who played first-person shooters *and* tended to default to stimulus-response navigation saw a measurable shrinkage in their hippocampus; those who defaulted to cognitive mapping actually increased the size of their hippocampus. The game itself wasn't necessarily the problem; it was how they navigated its virtual world.

The fact that a few months of playing video games produced measurable brain changes is surprising; the idea that these changes will cause measurable harm is more controversial. A smaller hippocampus is *associated* with various health conditions, but that doesn't mean it *causes* them. Still, there are some suggestive hints. Schizophrenia patients, for example, already have an unusually small hippocampus when they have their first episode of psychosis. Adolescents who have a family history of depression but don't (yet) have it themselves also have a smaller hippocampus. Even more telling, a 2002 Harvard study imaged the brains of forty combat veterans, roughly half of whom had PTSD, and also imaged the brains of their identical twins who hadn't experienced combat. The combat vets with PTSD had smaller hippocampi—but so did their noncombatant twins. That suggests the small hippocampus was present before they went into combat, and may be one of the factors that left them more vulnerable to PTSD.

Both hippocampal and caudate nucleus styles of thinking are important. You don't want to invent new ways of tying your shoes every morning and blaze new trails to your office on every commute. Routine, stimulus-driven learning is fast and efficient, and there's a reason we gravitate more and more toward it as we age. "This is a biologically adaptive mechanism," Bohbot says. "If you're a nomad, it frees you up to learn new things." But we need a balance of both systems, given the possible cognitive risks of letting the hippocampus atrophy. Bohbot's work suggests a self-reinforcing cycle: the more you rely on your caudate nucleus, the stronger it gets at the expense of the hippocampus, making it easier to default to it next time. "You can't get away with stimulating your caudate nucleus all day long and not have a cost to it," she says.

An obvious antidote, Bohbot says, is to explore more. "When you're exploring, you're going to make errors. And when you're making errors, that means you'll pay attention to your environment in order to find your way, so you're going to stimulate your hippocampus." In other words, don't be afraid to turn off the GPS and get lost. Technology, and progress in general, have given us superpowers: it no longer takes three years, or even eighty days, to circumnavigate Earth. The Concorde did it in under thirty-two hours in 1995; the International Space Station does it every ninety-two minutes. But sometimes, Bohbot says, it's worth taking the scenic route instead—especially if you're not sure where it leads.

In late December 1889, somewhere on the South China Sea between Singapore and Hong Kong, Nellie Bly and Elizabeth Bisland passed each other heading in opposite directions. By this time, both women had settled into the rhythm of traveling. "To sit on a quiet deck, to have a star-lit sky the only light above or about, to hear the water kissing the prow of the ship, is, to me, paradise," Bly wrote as she approached Hong Kong. Despite her boss's fears, she hadn't needed a male protector or missed any connections because she had too much luggage—though she did acquire an ill-tempered monkey in Singapore, which made transfers a little more complicated. She and Bisland were successfully navigating shifting schedules and local logistics in a bewildering series of languages and cultures, at a time when the far corners of the world were more diverse and less familiar than they are today. They were probably somewhere near the sweet spot, if not on the far side, of their personal Wundt Curves.

The itineraries that Thomas Cook & Son had hastily plotted for both women back in New York had been largely abandoned en route. But the company had one more role to play in determining the outcome of the race—if only indirectly. On January 16, 1890, Bisland caught the express mail train from Brindisi, at the southeastern tip of Italy, toward France—the same train that Bly had taken in the opposite direction two months earlier. If she could make it to Le Havre by six in the morning on the eighteenth, she would be able to cross the Atlantic on *Le Champagne* and complete her journey in as few as seventy-two days. But at the Paris suburb of Villeneuve-Saint-Georges, where she was to change trains for Le Havre, a Thomas Cook agent was there waiting for her with bad news: she was too late to catch *Le Champagne*.

Bisland reembarked on the mail train with a revised plan: she would continue on to Calais, where she would catch a ferry to Dover, take a train to London, then another train to Southampton where the *Ems* would take her across the Atlantic. But that plan too fell through: in London, she discovered that the *Ems* had been delayed until later in the week. So she took a night train to Wales and caught a ferry to Ireland, where she hoped to catch the *Etruria*—but it was pulled out of service and replaced with the *Bothnia*, among the slowest steamers in the Cunard fleet. At this point, Bisland was out of options. Her hopes of beating Bly were over, and even her chances of getting home in eighty days were in serious jeopardy.

Worst of all, *Le Champagne* had, in fact, waited for Bisland. The owner of Bisland's magazine had agreed to pay the staggering price of $2,000 to the steamer line in exchange for holding the ship for an hour or two; he cabled American diplomats in Paris to get them to obtain permission from the French government for the delay; he'd chartered a special train for $300 to rush Bisland from Villeneuve-Saint-Georges to Le Havre. But Bisland, thanks to the false information from the man claiming to be from Thomas Cook & Son, hadn't shown up. No one has ever figured out who he was or why he was there, though rumors inevitably swirled.

Bly, meanwhile, crossed the Pacific and made it to San Francisco on January 21, only to discover that a massive snowfall had closed all railway lines across the Rockies. *The World* had sent a reporter to meet Bly and chaperone her across North America, but his train got stranded in the Sierra Nevada. After five days, he received a telegram that Bly's ship was due the next day, so he hired a local miner to lead him overnight on skis, over accumulated snow that was reportedly as much as thirty feet deep on the summits, fifteen miles to the next unblocked train station. He was able to join Bly on a special train the newspaper had chartered, dodging the snow blockade by taking her south past Bakersfield, around the southern end of the Rockies, then east toward Albuquerque. The rest of the trip, back on the familiar rail lines of America, was a prolonged victory lap with cheering crowds at every stop. Bly made it back to her starting point in New Jersey in seventy-two days, six hours, eleven minutes, and fourteen seconds. Bisland arrived a few days later to considerably less fanfare, her trip having taken seventy-six days, sixteen hours, and ten minutes.

There were still blank spots on the map in 1890. No one had ever visited the North Pole, or the South Pole, or the top of Everest. But the successful conclusion of Bly's and Bisland's races signaled a change. To girdle the globe, you no longer needed to be a swashbuckling adventurer like Ferdinand Magellan, or fictional character like Phileas Fogg. You just needed to visit your local Thomas Cook office and buy a ticket. From a purely geographical perspective, the great age of exploration was drawing to a close. To keep scratching that itch, we would need to look elsewhere.

10

REDISCOVERING PLAY

The game was called "explorers." It was a Sunday afternoon, and six-year-old Cody Sheehy and his older sister, Carrie, were on a family picnic in the rolling foothills of the Blue Mountains in northeastern Oregon, a ninety-minute drive from their home. They picked a tree on the far side of the meadow, away from where their mother and a friend were sitting around a fire. Then they turned in opposite directions from each other and walked off into the woods. The goal was as hazily defined as the best childhood games are: experience the frisson of being alone in the forest; find something cool or unexpected; then return to the starting point to share your discovery.

But the return journey didn't go as planned. When Cody reemerged from the forest, his tree wasn't there. Neither was his sister, or his mother, or the campfire. It was a different meadow. He headed back into the forest, tried another direction, and found another meadow. It wasn't the right one either. He kept searching, his footprints crisscrossing the patches of snow lingering in clearings among the ponderosa pines. Eventually, he had to admit to himself that he was lost. It was 2:30 p.m. on a late April afternoon, with temperatures dropping toward freezing and clouds threatening rain. He saw a dirt road in the distance, and figured it must lead somewhere. He decided to follow it.

The purpose of play has puzzled big thinkers for millennia. Why do humans, along with mammals, birds, and other animals, engage in an activity that has no apparent benefit for survival or reproduction, but nonetheless burns energy and can get you lost in the woods? Plato saw childhood play as a form of practice that develops skills we'll need as adults. Jean-Jacques Rousseau figured that playing in nature civilizes us. Immanuel Kant thought that it liberates the spirit, and Friedrich Schiller thought that it releases pent-up energy. Modern biologists hypothesize that it helps develop the balance needed for high-speed chases, or prepares you for the emotions you feel after losing an aggressive interaction, or even teaches you about fairness and the consequences of cheating.

Maybe all these things are true, because play is a catch-all term for a wide variety of behaviors, from peekaboo to professional football. The media theorist Stephen Johnson, in a book on the creative power of play, defines it loosely as "all the things we don't *have* to do." And indeed, most of the attempts to define play scientifically converge on a crucial point: play is self-chosen, self-directed, and intrinsically motivated. Whatever its downstream benefits or ultimate evolutionary purpose may be, we do it in the moment because it's fun. "Happiness is never better exhibited than by young animals, such as puppies, kittens, lambs, &c., when playing together, like our own children," Charles Darwin wrote in *The Descent of Man*.

Exploring and playing aren't the same thing, but they're intimately connected. The former, in Marc Malmdorf Andersen's framework, is slope-chasing: seeking just the right dose of uncertainty to enable you to generate and resolve prediction error as rapidly as possible. The latter is slope-building: deliberately choosing the rules of your game to create new uncertainty and then resolve it. Will this block make the tower fall down? Can I hit the fastball? Can I find my way back to this tree? Finding that optimal slope is a moving target. The playground equipment in an unfamiliar park may initially offer plenty of uncertainty, Andersen says, "but it doesn't take many slides down the slide before kids find it more fun to climb up."

This link between playing and exploring—between slope-building and slope-chasing—has important implications. It suggests that the most productive exploring is also the most fun, fun being the intrinsic neural signal that we're maximizing our learning rate. It also suggests that the decline of play, both within our lifespans as we become curmudgeonly adults and across generations as we tether our children to low-risk activities and high-tech screens, has serious consequences. Scores on standard tests of creativity, for example, have been declining since the 1990s. To reverse these trends, we'll need to do better at heeding our most playful and exploratory instincts.

In 2007, the *Daily Mail* published an article about the links between health and time spent outdoors in nature. Accompanying the article was a striking visual that highlighted the steep decline of free play over the course of four generations of a single family in Sheffield. The great-grandfather, George Thomas, was free to wander on his own for six miles to a fishing hole in neighboring Rother Valley when he was eight years old in 1926. The family was too poor to afford tram fare or a bicycle, so he walked. A generation later, in 1950, eight-year-old Jack was allowed to roam on his own for one mile from the family home to play in a nearby forest. Jack's daughter Vicky, who was eight in 1976, could stroll to the swimming pool a half-mile away, walked to school on her own, and spent her free time riding her bike and playing unsupervised with friends in the park.

The trend reaches its grim conclusion with George's great-grandson Ed, who was eight when the article was published. He was allowed to walk to the end of his street, a mere three hundred yards away—but he didn't do that very much, because none of his friends played outside. His mother drove him to school in the mornings, then shuttled him to his piano lessons and various sports activities. On weekends, the family would sometimes drive to the countryside for a bike ride, but Ed didn't bike around his own neighborhood. "Traffic is an important consideration, as is fear of abduction, but I'm not sure whether that's real or perceived," his mother admitted. "Over four generations our family is poles apart in terms of affluence. But I'm not sure our lives are any richer."

It's not just the Thomas family. The widespread and dramatic decline of children's "home ranges," the swath of outdoor territory around where they live in which they can travel and play and explore independently, has been sparking concern among researchers for several decades now. Kids who roam widely around their neighborhoods spend more time outside and are more physically active, pick up stronger navigation skills, have greater knowledge of their surroundings, and may even end up less anxious as adults. But in countries around the world, the size of that home range has been shrinking steadily.

Of course, childhood wandering can be risky. Ed Cornell, a developmental psychologist at the University of Alberta, got his start studying home ranges in the early 1980s after getting a phone call from the Royal Canadian Mounted Police. They were searching for a nine-year-old boy who had gone missing after wandering away from a campsite. The officer wanted to know how far nine-year-olds typically roam. Cornell had no idea. He and a colleague, Donald Heth, quickly dug into the literature on the topic, but found slim pickings. They relayed their findings back to the police. "Well, that's not much," the officer admitted. "Don't worry, doc, we may get a psychic out here today."

Chastened by their inability to help, Cornell and Heth set out to answer the question by asking children to lead them on adventures to the farthest place they had ever traveled alone. "We followed them everywhere, including shortcuts through shopping malls, across snow-filled vacant lots, and once through an ongoing soccer game," they later reported. "Children interrupted their walks to throw stones, to stand on a fire hydrant to survey the upcoming path, or to dash off to kick a pile of leaves." The destinations, when they eventually reached them, turned out to be farther away than anyone had suspected. Six-year-olds, for example, ventured to spots an average of 769 meters away as the crow flies, three to four times farther than the distances reported in earlier studies that relied on interviewing kids instead of following them.

One of the metrics Cornell and Heth calculated was the angular dispersion of their young subjects' paths. If they traveled in a straight line from their home to their destination, the angular dispersion was zero. No one did that, of course. The more detours a child took from side to side, the greater the angle between two lines bracketing her entire route on a map. At seven years of age, the average angle was 138 degrees. At ten, even though the older children were presumably more competent navigators, the angle was 216 degrees. An angle greater than 180 degrees means that at least part of your route is heading in the complete opposite direction, away from your destination. "Independent children often go out of their way to take 'shortcuts' that are frequently longer and more hazardous than the original routes that they know," Cornell and Heth noted. Put another way, they're more interested in exploring than in getting from A to B.

These findings on home ranges and roaming patterns joined a growing body of knowledge about how people behave when they're lost. Psychologist Kenneth Hill, in his 1999 book *Lost Person Behavior*, identified nine main strategies that people use, including random traveling (what it sounds like) and using folk wisdom (following frequently misleading advice like "all streams lead to civilization"). The one that search and rescue experts recommend is to stay put—or, in the slogan of an educational program established after a lost nine-year-old boy died of hypothermia on Palomar Mountain in California in 1981, to "hug a tree." Another is direction sampling, which involves short forays in different directions from an anchor point. That was Cody Sheehy's initial response when he couldn't find the meadow where his mom was picnicking. A third strategy is view enhancing, which involves seeking high points or climbing trees to scan the surroundings. And then there's trail running, which means following the most obvious path, even if it seems to be leading in the wrong direction. That was Sheehy's next move, following the dirt road he spotted in the distance.

By the time Sheehy's family realized he was truly missing and called for help, it was late afternoon. They mobilized dozens of people, some on foot, some on horseback, some in trucks. They focused initially on an area around Deerings Meadow, where the family had been picnicking. But they underestimated his range. He was already miles away, following the dirt road that he hoped would lead him back down to the bottom of the Wallowa Valley, where his family lived.

This was 1986, a time when the typical home range of six-year-olds had already begun to plummet. By one estimate, children's unstructured playtime dropped by 25 percent between 1981 and 1997. But

Sheehy lived on a ranch and had plenty of freedom to roam. His father was an agricultural consultant and former marine who, the previous year, had taken his family with him for a work contract on the grasslands of Inner Mongolia, leading his three children on hikes of up to eight miles. He was the type of kid who climbed trees, caught crayfish, and played explorers in the woods. "Do you think a six-year-old can walk more than five miles?" a police officer asked Sheehy's mother skeptically as the search dragged on and its perimeter widened. "This one can," she replied.

To Cornell and Heth, there's a clear link between the skills you pick up playing freely and the skills you need to navigate the unknown. "Adventure fosters adjusting to unanticipated and changing circumstances," they write. By sunset, Sheehy had already traveled somewhere between six and nine miles. He slipped into a creek at the bottom of a steep gulch, soaking his legs, but kept walking. The temperature was close to freezing now, and his feet began to hurt. A few hours later, he came to an abandoned white house, which struck him as spooky. Two coyotes emerged from the darkness and moved toward him. He fled and climbed a tree, where he wedged himself onto a branch and, eventually, dozed. But the cold was gripping his body now, so he couldn't stay asleep. He climbed back down and trudged onward.

That play is a rehearsal for real life is a familiar and uncontroversial idea. Pouncing on a ball of yarn will hone the skills that enable you to later pounce on a mouse; finding your way through the woods as a game will prepare you for finding your way through the woods when you're actually lost. This assumes that play is rational, maximizing some future outcome. But it's not—at least, not always.

To be sure, there's plenty of evidence that some types of exploratory play are carefully tuned to maximize information gain. Even infants gaze longer at events that violate their expectations, and as they get older their play choices gravitate toward options with higher uncertainty and prediction error. In one study, for example, thirteen-month-olds were shown two boxes containing a mix of differently colored balls, from which the experimenter extracted either a series of balls of various colors (as you'd expect) or a series of balls that were all the same color (a puzzling and unexpected pattern). When the toddlers were released from their parents' laps, they were more likely to crawl toward the surprising (and thus information-rich) box and play with its contents. In older children the pattern is even more pronounced: play is more fun when you're discovering something new or unexpected.

Outside the laboratory, though, play doesn't always follow those rules. "Even in the context of relatively straightforward exploratory play, children tend to violate principles of efficient planning and rational action," write MIT researchers Junyi Chu and Laura Schulz. That may seem obvious when children say "Let's stick crayons into the vent," but it's just as true, they point out, when adults say "Let's use black and white stones to control territories on a grid." In a 2023 study, Chu and Schulz assigned various tasks to four- and five-year-olds: find a ball hidden in a set of drawers, retrieve a pencil Velcroed partway up a wall, figure out which button makes a robot dance. When the children were simply asked to complete the task, they did it efficiently. When the request was framed as play, on the other hand—"We're going to play a hide-and-seek game with this ball. I'm going to hide it in a drawer, and you can look for it"—the children chose deliberately inefficient strategies. They would jump for the pencil farthest out of reach, or start their search for the ball in the least-likely drawers. They knew the "right" way to fulfill the task, but that didn't seem as fun.

Play, Chu and Schulz conclude, involves "incurring seemingly unnecessary costs to achieve arbitrary rewards." That echoes the philosopher Bernard Suits's definition of games, which he outlined in a 1978 book called *The Grasshopper*: playing a game, Suits wrote, is "a voluntary attempt to overcome unnecessary obstacles." Suits's book is written as a parable featuring the grasshopper from Aesop's fable, who has played and danced the summer away and now faces starvation as winter approaches. But this grasshopper makes no apology: he argues instead that playing games is our highest calling, since in a hypothetical utopia where all material wants are satisfied that's all we would do.

We'll come back to Suits's seemingly hyperbolic claim about the primacy of play later, but there are other reasons to see play as a worthy pursuit even if you don't think it's worth dying for. One theory, according to Chu, is that "play solves a meta exploration/exploitation problem"—that is, it helps us to explore different ways of exploring. When the children in their studies start searching for the ball in the least-likely drawer, they're failing to gather information efficiently. But they're experimenting with different ways of solving problems that may eventually transfer to other as-yet-unforeseen contexts. "Liberated from any practical goals—even the goal of learning—play may be a means of increasing innovation," she and Schulz write. "Our capacity to invent and solve small problems, and to find it rewarding, may help human learners solve a big problem: the problem of how to generate new ideas and plans in an infinite search space."

Kids may also be tapping into another important source of uncertainty: themselves. "Maybe play is motivated by learning about our own abilities, preferences, and knowledge, not just how the external world works," Chu told me. Jumping for the most out-of-reach pencil isn't the easiest way to get the pencil, but it is a way of finding out whether you can jump that high. That echoes a point made by George Loewenstein in a famous paper titled "Because It Is There: The Challenge of Mountaineering . . . for Utility Theory." Why do people embark on a pursuit that mostly consists of "long periods of stultifying boredom punctuated by brief periods of terror," he asked? Among other ideas (including, facetiously, the proposal that "meaning-making may also be enhanced by loss of body parts"), he suggested that self-signaling plays a role: you find out, and demonstrate to yourself, who you really are and what you're capable of.

Some of history's most innovative people, Marc Malmdorf Andersen points out, were known to be famously playful. Mozart loved playing music backward, for example. He once composed a canon in F major called "Difficile Lectu" with words in fake Latin carefully chosen so that, when sung with the strong Bavarian accent of his lead tenor, it would sound like he was singing "Lick me in the ass." Alexander Fleming, who discovered penicillin, had a simple explanation for his scientific process: "I play with microbes. There are, of course, many rules in this play, and a certain amount of knowledge is required before you can fully enjoy the game, but, when you have acquired knowledge and experience, it is very pleasant to break the rules and to be able to find something that nobody had thought of."

Richard Feynman, the physicist who helped kickstart the field of quantum computing, had a playful streak that bordered on notoriety. "Feynman is the young American professor, half genius and half buffoon, who keeps all physicists and their children amused with his effervescent vitality," the British physicist Freeman Dyson wrote to his parents in 1948. Dyson later revised his opinion: Feynman was, in fact, "all genius and all buffoon." His memoirs are packed with tales of his unconventional amusements: stealing doors from his undergraduate fraternity, picking the locks of top-secret filing cabinets in Los Alamos during World War II, playing bongos for a modern ballet that tours to Paris. As a grad student at Princeton, he conducted elaborate navigational experiments on the ants raiding his kitchen, and accidentally blew up a giant glass carboy in the cyclotron lab while trying to settle an argument about which way a lawn sprinkler would rotate if it were underwater.

Feynman's intellectual curiosity wasn't confined to physics. His foundational insight about quantum computing came from dabbling in the unfamiliar world of computer science—but that was far from his first foray outside of physics. In the dining room at Princeton, instead of sitting with his fellow physicists, he would rotate for a few weeks at a time through the tables of various other specialties—philosophy, biology, math—trying to understand the problems they were grappling with. He sometimes got so interested that he ended up attending their classes for a semester. As a physics professor, he spent a summer vacation volunteering in a biology lab and later spent a sabbatical year in a different biology lab. He explored out-of-body experiences in sensory deprivation tanks, and lectured publicly on the mathematics of ancient Mayan manuscripts. He also promised to teach science to an artist friend if the friend would teach him how to draw—and ended up becoming adept enough to host a one-man show and get a commission to provide wall-art for a local massage parlor.

None of these hijinks would be interesting to posterity if Feynman hadn't also been a brilliant physicist. But in his view, the two facets—the genius and the buffoon—were inextricably linked. After spending the war years in Los Alamos working feverishly on the development of the atomic bomb, Feynman started as a professor at Cornell University in 1946. He liked Ithaca and enjoyed teaching, but couldn't get anywhere with his research. He would write a sentence or two about gamma rays or some other problem, then run into a brick wall. After the stress of the Manhattan Project, he concluded, he was burned out, perhaps for good. This dry spell dragged on, sapping his morale and self-confidence, until one day he had an epiphany:

> Physics disgusts me a little now [he told himself], but I used to enjoy doing physics. Why did I enjoy it? I used to play with it. I used to do whatever I felt like doing—it didn't have to do with whether it was important for the development of nuclear physics, but whether it was interesting and amusing for me to play with.

Within a week of this realization, he noticed someone tossing a plate in the air in the cafeteria. Idly, watching the red emblem of Cornell spinning around, he observed that the plate was rotating faster than it was wobbling. Out of curiosity, he decided to calculate the motion of a spinning plate, and out of the complex equations emerged a simple ratio: two rotations for every wobble. He shared the result with a senior colleague, who asked him why it mattered. "Hah!" Feynman responded. "There's no importance whatsoever. I'm just doing it for the fun of it."

But he kept fiddling with the wobble equations, which led him to think about how electrons orbit the atom, and how the rules of special relativity affect those orbits. One thing led to another. "It was effortless. It was easy to play with those things," he later recalled. "There was no importance to what I was doing, but ultimately there was. The diagrams and the whole business that I got the Nobel Prize for came from that piddling around with the wobbling plate."

Creativity is not the same as intelligence: you can have a high IQ and low creativity, or a low IQ and high creativity. In fact, on a societal level, those two traits seem to be moving in opposite directions. In the early 1980s, an American-born researcher in New Zealand named James Flynn noticed a pattern in the results of IQ tests from around the world: we're getting smarter. The Flynn effect, which has since been confirmed in data from dozens of countries around the world, shows that people have been gaining the equivalent of about three IQ points per decade for over a century. The reasons for this change are still being debated, but Flynn believed that modern society demands—and thus nurtures—abstract thinking in a way that even our recent ancestors didn't need.

The closest equivalent to an IQ test for creativity is called the Torrance Tests of Creative Thinking, devised by a psychologist at the University of Minnesota named Paul Torrance in the late 1950s. It involves a series of verbal tasks such as coming up with unusual uses for everyday objects like tin cans and bricks, and nonverbal tasks like completing abstract unfinished drawings. The tests have been administered to tens of thousands of schoolchildren, and every decade or so the scores are renormalized by Scholastic Testing Services to ensure that the results are comparable across generations, just like IQ and SAT tests.

In 2011, an education professor at the College of William & Mary named Kyung Hee Kim published an article in the *Creativity Research Journal* about what she dubbed the "creativity crisis." By analyzing the renormalization data from Torrance tests dating back as far as 1966, she found that levels of creativity in American children and adults had remained relatively steady until about 1990, and then began declining. The drop was precipitous: as much as a full standard deviation, meaning that 85 percent of children in 2008 were less creative than the average child in 1990. Even worse, a follow-up paper published in 2021 found that the rate of decline "significantly escalated" after 2008. The Flynn effect says we're getting smarter, but the Torrance data says we're getting less imaginative and innovative. That's a problem, because by one estimate, the Torrance Tests predict adult creative achievement three times more accurately than IQ tests.

It's probably fair to say that innovation is one of the many traits that every generation grumpily worries that its kids are losing. Still, there are scraps of real-world evidence that back up Kim's observation. In 2023, for example, University of Minnesota researchers published an analysis of 45 million scientific papers dating back to 1946, along with 3.9 million U.S. patents starting in 1976. For each one, they assigned a "disruptiveness score." Highly disruptive papers altered the entire trajectory of their fields, so that earlier papers on the same topic would no longer be cited by later research. At the other end of the spectrum, highly consolidating papers confirmed or elaborated on existing theories. A consolidating paper might be important enough to win a Nobel Prize, but it would add to earlier papers rather than displacing them.

Sure enough, the analysis showed that scientific and technological research have been getting steadily less disruptive in recent decades. You might wonder whether we've simply grabbed all the low-hanging fruit in fields like physics, so there's not much left to discover. Moore's Law, for example, used to reliably predict that microchips would double in speed every eighteen months or so, but chips are now so small that they're bumping up against fundamental limits imposed by quantum mechanics. That can't be the whole answer, though. Disruptiveness scores declined by similar amounts on similar timescales across life sciences, physical sciences, social sciences, and technology. That suggests it's not a subject-specific problem but rather a more general societal issue.

Kim grew up in South Korea and worked there as a teacher for ten years before moving to the United States for graduate school at the University of Georgia's Torrance Center for Creativity and Talent Development. One of the factors contributing to the decline of creativity in the United States, she believes, is a growing shift toward the sort of test-focused, rote-memorization-heavy education that she experienced in South Korea. That may be one of the reasons free play dropped by 25 percent in the 1980s and 1990s. Kids need time to think, reflect, and explore on their own terms. Problem-solving skills are important, but so are problem-*finding* skills. One study found that adolescents came up with more creative responses to challenges they created themselves than to those a researcher presented to them—an echo of Marc Malmdorf Andersen's definition of play as "slope-building."

Halley's Comet was making one of its periodic flybys in the spring of 1986. As Cody Sheehy hiked onward through the drizzly night, he distracted himself by scanning the skies for the passing fireball, which had made headlines with its closest approach just a few weeks earlier. He didn't spot it, and his feet were growing increasingly painful, but he kept going. He would walk for "a thousand minutes" (as he later told reporters), take a half-hour break, then repeat. He knew his home, on the valley floor, must be downhill from where he'd gotten lost. But the rolling hills made it difficult to tell whether he was heading in the right direction, especially in the dark. Shortly after 6:00 a.m. he finally got a view off the edge of the plateau as the sun rose. There below him, in the distance, were houses and roads. He still didn't recognize the location, and the town was a long way away. But the end, finally, was in sight.

These days, kids in the United States average about six hours a day of screen time by the time they're eight years old, according to the Centers for Disease Control and Prevention. Few of these kids would be able to match Sheehy's marathon hike. Of course, they wouldn't get lost in the first place, since they'd be safely tucked away in their rooms watching television rather than roaming in the woods. But their safety comes at a cost. "I connect technology more to passive play," Kyung Hee Kim told an interviewer. "So instead of going outside where you actively play, you explore outdoors, and you get hurt sometimes, and then you learn something—instead of children doing that, nowadays, you sit in front of an iPhone or TV or computer. Passive play is really hurting your creativity."

Screen time has certainly contributed to the decline of active outdoor play, but it's just one part of the story. We also have to consider the other side of the equation. "It's not only the case that technology has come along and now all the kids only want to look at their phones," Andersen says. "It's also the case that they can't go to the park by themselves anymore." And if they do, a parent will come along to supervise, and the playground will have been de-risked, with any potentially dangerous—or, from a kid's point of view, fun—pieces of equipment removed. Screens are all the more attractive because exploring the real world is more boring than it used to be.

"I sometimes get these questions—you know, should we get people to play more?" Andersen told me. "But it doesn't make any sense to put it like that." The crux of Andersen's theory of play as slope-building is that it's an internally motivated behavior. Play is what happens when we're permitted to follow what he calls a "hedonistic principle of curiosity." It's what we want to do, not what we ought to do. We may have strong ideas about what types of play will be good for kids, but all we can do is create opportunities and let them decide how to use them. For his own kids, Andersen says, "I try as much as I can to support their ideas and the things they want to do. I try to say yes a lot, in other words. If they think it's worth pursuing, then it by default is worth pursuing for them."

The same principle applies, more or less, for grown-ups. "Adults would play more if they could," he says. "It's the behavior that humans will gravitate toward when they are fit and taken care of and not stressed and all that stuff." For some people, that means creating space outside their work and family responsibilities to follow the hedonistic principle; for others, it means finding opportunities to play within the context of work. There's isn't one right answer; nor, as Richard Feynman found, does there have to be a clear distinction between what is work and what is play. "Play has this innovative potential, especially if experts play within their given domain," Andersen says. "If someone like Mozart wants to surprise himself, what would he do? And what would it mean for the rest of us?"

At eight thirty in the morning, Sheehy reached the outskirts of Wallowa, a town of about eight hundred people. He'd been walking for nearly eighteen hours, and had covered somewhere between fourteen and twenty miles. He was determined to walk all the way home, so he knocked on the door of a nearby house to ask how far it was to his grandfather's place. The woman inside called the sheriff, who sent someone to drive Sheehy home. The story became a minor national media sensation. His feet were badly swollen, with tendinitis that kept him on crutches for days, but he emerged otherwise unscathed. He's now a documentary filmmaker in Tucson, a second career after training as an engineer. One of his films, about the water crisis in the Southwest, picked up a regional Emmy. He and his wife own a fifty-four-foot sailboat, which they sail down the coast to Mexico each year. They've gone as far as Nicaragua. "You can get to the most rugged, remote wilderness, a hundred miles from the closest village, or have an island all to yourself," he told environmental writer Emma Marris when the two of them retraced his route through the woods three decades later for *Outside* magazine.

Not every story of a lost child ends happily. Exploratory play, more or less by definition, involves risk, which is a big reason why play has both declined and taken on more passive forms over the decades. For Sheehy, though, there are no regrets. His ordeal was hard—but, as we'll see in the next chapter, that's what gave it so much meaning. "Over the course of your life, you push through a lot of physical barriers," he told Marris. "As you grow older, your first coach helps you break through barriers, and maybe in the military you learn to push through barriers or maybe in your first hard job. As a little kid, I had this opportunity to be tested and learn that there really aren't any barriers. I think a lot of people figure that out. They just might not figure it out at six."

11

THE EFFORT PARADOX

It was, the *New York Times* later opined, "one of the greatest relay races the carnival has produced." The brand-new upper deck at the University of Pennsylvania's Franklin Field was packed to capacity for the 1926 Penn Relays, with forty thousand spectators on hand for what was already the largest and longest-running track meet in the United States. A newly installed loudspeaker system, replacing the outdated and largely ineffective megaphones used in previous years, kept fans abreast of the latest developments and whipped them into a frenzy during close races. And the two-mile college relay was set to be the highlight of the final day.

The favorites were the defending champions from Georgetown University, who had set a world record of 7:42 at the previous year's Penn Relays. But strong challenges were anticipated from Boston College—and, unexpectedly, from Columbia University in New York. The previous day, the unheralded Columbia quartet had upset Georgetown in the sprint medley relay. In that race, Georgetown led by three yards heading into the final 880-yard leg, but Columbia's anchor runner, twenty-two-year-old Joe Campbell, had quickly surged into the lead then, as the *Times* reported, "fought off two desperate challenges by Eddie Swinburne, Georgetown's anchor man, and beat the Blue and Gray to the wire by a scant foot." Columbia's overall time for the sprint medley and Campbell's individual time of 1:53 for the final 880 yards were both just one second off the respective world records.

Campbell's heroics in the sprint medley, which made headlines in newspapers across the country, were a pleasant surprise for the generally hapless Columbia team. Campbell had graduated with a B.A. the previous spring and left academia behind to start work at his father's wholesale clothing store. But the novelty of the real world—and, more specifically, of folding endless pairs of socks and stockings for display—had worn off within a few months. "Business, as I have seen it so far," he wrote in his journal, "reduces living men to dull machines, that go on from day to day working at stupid tasks with not the slightest idea of what they are working for."

Rather than persist in the sock business, Campbell quit his job (to the apparent relief of his father, who hadn't appreciated his son constantly quizzing his fellow employees about the pointlessness of their existence). He reenrolled at Columbia as a graduate student for the spring semester in 1926 and was now running track while working on a master's thesis about the Arthurian legends surrounding the quest for the Holy Grail. His approach was unconventional, seeking to mine the ancient tales for psychological insights rather than simply parse their historical development. "Modern questers for secrets of the grail swarm along a well worn route, and battle their ways through the multitude around them," he wrote in the introduction to his thesis. "The path which I propose to follow lies aloof from the main road, and no one has ever pressed it to conclusion."

It was on Campbell, then, that Columbia's hopes in the marquee two-mile relay rested. The race didn't start well, though. Boston College's first runner established an early lead, and their second leg widened the gap to two full seconds over the chase pack. In the third leg, Columbia's Johnny Theobald tried to claw back some ground, but Boston College's lead at the final hand-off was still substantial: the *Times* had it as 12 yards; the *Allentown Morning Call*'s reporter thought it was more like 20 yards; the *Chicago Tribune* put it at 35 yards. Campbell himself, six decades later, remembered it as 30 yards.

Whatever the exact margin, Campbell knew what he had to do in his final 880-yard leg, and what it would require. One of the books he'd acquired for his thesis work was a volume containing the writings of Hartmann von Aue, a twelfth-century poet and knight who introduced Arthurian legends into German literature. Among the notes

Campbell scrawled in thin pencil in its margins is this: "What I dared to suffer / you dare not endure."

Exploring, as I've described it in these pages, seems to have two very different meanings. On the one hand, it's an arduous journey into the unknown where suffering is inevitable and failure looms as an ever-present possibility, as celebrated in the perpetually bestselling (and sometimes posthumous) accounts of famous mountaineers and polar explorers. On the other hand, as I argued in the last chapter, it's a joyful act of discovery in which new insights about ourselves and the world around us are obtained by maintaining a playful spirit and an open mind. My goal in this chapter is to argue that there is no contradiction between these two versions of exploring—that the toil and strife of the exploring bros doesn't cancel out the joy of discovery, but instead heightens it. True exploring, even of the armchair variety, is hard, and that's a big part of what we love about it. Joe—or, as he's better remembered, Joseph—Campbell, Columbia's ace half-miler, went on to achieve lasting renown as the twentieth century's most influential interpreter of mythology. He argued that cultures from around the world tell similar stories drawing from an archetypal "monomyth" that he dubbed the hero's journey, imprinted in the collective unconscious of our species. The ritualized substages of this journey—a call to adventure, the initial refusal of the call, a meeting with a mentor, an ordeal, and so on—can be divided into three acts: departure from the familiar world, a challenging adventure in an unfamiliar world, then a return to the familiar. In other words, it's a tale of exploration.

Campbell's framework eventually permeated pop culture. He lunched with an admiring Bob Dylan, gave talks alongside the Grateful Dead, and appeared onstage as a special guest at one of their concerts. ("They didn't know what they were saying, and we don't know what we're saying either, but we think we're saying the same thing," Jerry Garcia quipped about the parallels between ancient mythology and modern rock.) Most famously, Campbell became friends with George Lucas, who had reworked his draft of the first *Star Wars* script after reading Campbell's most influential book, *The Hero with a Thousand Faces*. Campbell eventually visited Skywalker Ranch for a marathon viewing of the original trilogy with Lucas—the first time that anyone had watched all three of the movies in a single day, according to Lucas. "You know," Campbell said afterward, "I thought real art stopped with Picasso, Joyce, and Mann. Now I know it hasn't."

These days, Campbell's hero's journey rubric is a staple of screenwriting and storytelling classes, familiar to the point of cliché. Campbell himself, by the time he died in 1987, had come to be seen as a Yoda-like figure of New Age enlightenment, dispensing encouraging slogans like "follow your bliss." This advice sounds a lot like Marc Malmdorf Andersen's hedonistic principle of play, and Campbell's own life, following his brief career in the sock business, lived up to the dictum.

As a college student, he paid his bills by playing saxophone at dances and events, played football at Columbia until a collision with eventual Hall of Famer Walter Koppisch broke his nose, and surfed in Hawaii with the Kahanamoku family. On a summer trip to Europe, he struck up a friendship with the Indian mystic Jiddu Krishnamurti on the transatlantic steamer; in Paris, he met the sculptor Antoine Bourdelle and posed for a bronze bust by one of Bourdelle's students. His introduction to running came during his sophomore year, when he had to run around the indoor track for phys ed. "I never had the ability to let someone be ahead of me," he later recalled. On the ten-laps-to-a-mile track, there was *always* someone ahead because he kept lapping people, so he kept going faster and faster. The track coach spotted him and invited him to join the team. By the following year, he was rooming with Olympic champion sprinter Jackson Scholz at the 1925 national championships in California, where he finished fourth in the 880 yards.

For all his adventures, though, Campbell didn't hesitate in later years when asked to identify his greatest experience: the 1926 Penn Relays. When he got the baton for the final leg of the two-mile relay, 12 or 20 or 35 yards behind Francis Daley of Boston College, he was leaving behind the comforts of ordinary life and entering the trackless internal wilderness familiar to everyone who has ever run a race at the edge of their abilities. What the outcome would be, and whether he would return bearing the prize he and his teammates sought, could only be revealed by plunging in headlong. "Perhaps these early experiences in athletic competition shaped Campbell's later concern with the hero as a mythic figure," one of his posthumous biographers wrote. "But for the time being, his concern was not the morphology or inner sense of the hero journey, but its literal enactment."

Little by little, Campbell closed the gap on Daley, drawing even at the beginning of the final straight then pulling away to win by ten yards. Columbia's second victory of the meet was the top story in the sports section of that evening's *New York Times*, with Campbell singled out for effusive praise. Six decades later, the memories were still fresh. "If anyone would ask me what the peaks were, the high moments of my life experience—really, *zing!* the whole thing in a nutshell—those races would be it," he told an interviewer. "More than anything else in my whole life."

Here's the thing about running a half-mile in under two minutes: it hurts. Physiologically, the distance lies right on the border between sprinting and endurance, so it's the longest sustainable sprint. But sports scientists have found that physical pain isn't actually what limits runners; instead, it's their sense of effort, a subjective feeling sometimes defined as "the struggle to continue against a mounting desire to stop" (or, less poetically, as "intensification of either mental or physical activity in the service of meeting some goal"). Effort is a broader and more amorphous concept than pain, and it's thought to serve as a means of homeostatic control: when we're pushing too hard, or squandering valuable resources, or making any sort of decision that our evolutionary wiring codes as dangerous or counterproductive, the perception of effort nudges us to back off.

The usual assumption is that effort is a negative: "toil and trouble," as Adam Smith called it in *The Wealth of Nations* back in 1776, subtracts from the value we assign to things. If you can purchase a coffee table that arrives in pieces with a bag of seemingly mismatched screws and some inscrutable pictographic instructions, or simply buy the same thing preassembled, economic theory predicts that you'll be willing to pay more for the latter. And it's not just about money. The law of least effort, formulated by the American psychologist Clark Hull in the 1940s, dictates that given two choices with similar outcomes, an organism will choose the option requiring the least effort.

Bizarrely, though, studies have found that we actually value the coffee table we've had to grapple with more highly than the identical preassembled version, a phenomenon now known as the IKEA effect. George Mallory wanted to climb Mount Everest "because it's there." We can speculate about his other motivations: reaching the highest point in the world, eternal fame, and so on. And we can assume that even modern-day Everest summiteers are partly motivated by the pursuit of bragging rights. But the fact remains than many of us head anonymously to the mountains, run midpack marathons, and do Sudoku puzzles—all activities that, like purchasing Swedish furniture, involve considerable unnecessary effort. The first marathon you run may be motivated by a desire to improve your health or by a Mallory-esque desire to find out what's on the other side. But the second one is likely fueled by something else.

All of these behaviors are examples of what University of Toronto social psychologist Michael Inzlicht has dubbed the effort paradox: sometimes we value experiences and outcomes (and coffee tables) precisely *because* they require effort, not in spite of that fact. Joseph Campbell's victories at the Penn Relays covered him in glory, but it wasn't the adulation of others that stuck with him through the decades. And it wasn't just those two races. *All* of his peak experiences came while running, he said near the end of his life, because "there is a kind of mystical bliss that comes when the body is overtaxed."

Inzlicht and his colleagues have come up with several possible explanations for why we find both physical and cognitive effort so satisfying. One is cognitive dissonance: if you do something that's really hard for an outcome that you don't consider particularly valuable, you suffer an unpleasant disconnect that you assuage by convincing yourself that the outcome was valuable after all. If I worked so hard to get this, I must really like it, you tell yourself. This theory makes sense in humans, but is less convincing in other species. If you train starlings to fly various distances to obtain identical color-coded treats, they'll end up liking the color of treats they had to fly farthest for, and that preference will remain, even when the treats are placed an equal distance away. The effect shows up in locusts too, which suggests that cognitive dissonance isn't the whole story, since locusts don't do a whole lot of introspecting about their motivations.

Another possibility is that rewards obtained from difficult tasks seem extra sweet because of the sharp contrast between the unpleasantness of working hard and the joy of achievement. As with cognitive dissonance, the contrast theory assumes that what we really value are the fruits of hard effort, rather than the effort itself. But that's not necessarily the case. The theory of learned industriousness assumes that over time we learn that working hard leads to rewards, so we begin to value the effort itself, like Pavlov's dogs salivating at the sound of a bell. Predictive processing too offers an explanation for why we might enjoy the process as much as the outcome: doing hard things gives us access to steeper slopes of uncertainty reduction, which in turn feels good. If you buy a coffee table, you've got a coffee table; if you assemble one, you also gain knowledge not only about how coffee tables are put together, but also about your own capabilities.

The thirst for challenge shows up even in the way children play. Researchers at Harvard and the University of California, Berkeley teamed up to test a group of five- to ten-year-old children playing a game they called beach bowling, which involved knocking over a set of pins by throwing either beanbags or batons at them. The children were allowed to choose various game settings like the weight of the pins and the throwing distance. There were two versions of the game: in one, they would win stickers if they knocked down all six pins; in the other, the goal was "to have as much fun as possible." Children who were invited to simply have fun consistently picked harder settings than those playing for a prize.

"When children play, they're not just trying to maximize extrinsic reward," Harvard's Elizabeth Bonawitz, one of the study's authors, explains. Their motivations are internal. It could be in service of learning about the world, as Marc Malmdorf Andersen and predictive processing theorists argue. Or more broadly, as Junyi Chu's work suggests, we might be learning how to learn and generating solutions to problems we haven't yet encountered. But there are other possible rewards that we might be pursuing when we voluntarily take on challenges, Bonawitz and her colleagues point out: mastery, autonomy, social connection, aesthetic experience—and even meaning, famously tricky though it is to define.

We don't have to precisely articulate the meaning of life in order to explore whether effort contributes to it. "If you ask people whether something is meaningful, they can answer," Inzlicht says, "but they use their own internal rubric to figure out what that means." Working with colleagues Aidan Campbell and Joanna Chung, Inzlicht has developed a ten-item Meaningfulness of Effort scale that asks people how strongly they agree with statements such as "When I push myself, what I'm doing feels important" and "Doing my best gives me a clear purpose in life." The scale captures differences not in whether people exert effort, but in how they view that effort. "You can imagine that some people are willing to work hard, but go about it from a sense of duty and responsibility," Inzlicht explains. "But other people—call them 'joyful workers'—this is what they live for. This is what gives them purpose. This is what makes them feel important. This is what helps them make the world make sense."

What's most interesting about the Meaningfulness of Effort scale is what it predicts. People who score highly tend to report greater levels of job and life satisfaction; they make more money and have higher-status jobs; they're happier (or in more technical terms, have greater subjective well-being). They have fewer mental health issues and—despite loving the grind—lower rates of burnout. Those findings remain true even when you control for other constructs like conscientiousness, which is one of the "Big Five" personality traits that psychologists use to classify people. There has been lots of debate in recent years over whether popular concepts like "grit" are just new names for old concepts. Meaningfulness of Effort is a subcomponent of conscientiousness, Inzlicht says, but it has distinct explanatory power. Willingness to exert effort is important, but how you *feel* about that effort also seems to matter.

How you score on the scale is partly born and partly learned, Inzlicht suspects. "Effort and reward are related in the real world," he says. "But for some people effort doesn't pay." If you're born in a poor neighborhood or face prejudice such that people don't reward you to the same extent as they reward other people, your effort-reward link will be different. Conversely, recent results from his lab suggest that if you consistently reward people for choosing the harder option, they'll gradually learn to make more effortful choices.

Strictly speaking, it's possible to "explore" by trying a new ice cream flavor, or by simply scrolling through your social media feeds. In each case, you could argue that you're making what Joseph Campbell called "a bold beginning of uncertain outcome." And it's true that the rewards are uncertain. But the stakes are laughably small. No effort is required, and there's no risk of failure.

You might think this sounds like an ideal version of exploring. Stimulate your reward centers by creating and then resolving uncertainty, but without the risk that you'll accidentally sail off the edge of the earth? What's not to like? And indeed, social media companies have built massive fortunes on the attractiveness of this model. But not many people would argue that it has made them happier or more fulfilled. One of Inzlicht's studies explores the links between digital switching—scrolling through YouTube or TikTok, say, and jumping from post to post—and boredom. When people are bored, they switch ever more rapidly; but greater switching paradoxically leads to worse boredom. The titillation of novelty wears off if you don't take a chance and invest your time and attention in something.

Campbell argued that we're wired to respond to a particular story arc, one that takes us from the known to the unknown and then—after great effort and many trials—back to the known. When I first heard that Campbell had been a world-class runner in the 1920s, my instinctive reaction was: "Of course!" Every time I go for a run, I pass through those stages. It's—*zing!*—the whole thing in a nutshell. Campbell's idea of a Jungian collective unconscious has fallen out of favor these days, and his comparative mythological work is outside the academic mainstream. But the arc itself endures—in the stories we tell, in the way we understand them, and more recently in the findings of modern neuroscience.

From 2011 to 2014, the Defense Advanced Research Projects Agency funded a multiuniversity research project called Narrative Networks. DARPA is the U.S. military's skunkworks agency for pursuing wild and crazy innovations. They were among the earliest funders of quantum computing research, for example. The goal of the Narrative Networks research was to figure out what happens in our brains when we hear a compelling narrative, and, conversely, what elements make a narrative compelling. The military hoped to figured out how to counter terrorist narratives with more convincing stories of their own—"To win wars not with weapons but with words," as two of the researchers involved put it.

One of the DARPA-funded teams was led by Paul Zak, who heads the Center for Neuroeconomic Studies at Claremont Graduate University, just outside Los Angeles. Zak started out with a PhD in economics but soon became dissatisfied with the field's unrealistic assumptions about human behavior. "You can build any model you want and predict some beautiful theorems, but it's not *useful*," he says. So he went to Harvard as a postdoctoral researcher to study brain imaging. These days, as a neuroeconomist, he looks for connections between what's happening in the brain and how people make choices in the real world.

Zak is best known for his research on oxytocin, the hormone associated with social bonding. His work on the neuroscience of narratives began with a study where viewers watched one of two short videos showing a father with his two-year-old son who has a soon-to-be-fatal brain tumor. One video featured a flat narrative structure, simply showing footage of the pair visiting the zoo; the other followed a hero's journey narrative as the father struggles to accept his son's diagnosis. The second one engaged viewers more, triggered higher levels of oxytocin in blood tests, and subsequently led them to behave more generously in a negotiation with strangers.

Spurred by DARPA's interest in field-deployable measures of engagement, Zak and his colleagues have refined their techniques. These days, they can use data collected by ordinary smartwatches to pick up subtle variations in heart rhythm that correlate with levels of oxytocin and other markers of engagement. Zak's spin-off company, Immersion Neuroscience, now works with Hollywood studios, ad agencies, and other storytellers to provide quick neuroscientific feedback on audience engagement. Zak points to data suggesting that more than half of Hollywood movies follow a hero's journey-type narrative, as do the majority of most-watched TED talks. "Structure is really important, because the brain is so lazy in terms of spending resources," he says. "If you don't compel me to remain immersed in the story, then it would rather just space out and idle because that saves energy."

Crucial to that structure is the struggle in the second act: what Campbell called the Road of Trials, and what later interpreters of his work sometimes call the Ordeal. Another DARPA-funded project, this one at Georgia Tech, ran brain imaging studies while subjects watched suspenseful films like Alfred Hitchcock's *North by Northwest* and *The Man Who Knew Too Much*. At these key moments, the researchers saw neural signatures of tunnel vision: brain areas responsible for peripheral vision shut down, and recall of plot-relevant details increased while less relevant details were forgotten. These moments are far more interesting to us than the eventual dénouement. We don't care what prize or secret or weapon the characters are pursuing, as long as *they* care enough to go through the Ordeal.

Hitchcock had a name for the plot devices that his characters chased: MacGuffins. The name comes from an anecdote he liked to tell about two Scotsmen on a train; a MacGuffin, it turns out, is a device for trapping the nonexistent lions of the Scottish Highlands. The technique reached its apotheosis in *North by Northwest*, in which Cary Grant pursues a nefarious agent played by James Mason, and in return is chased through a cornfield by a crop duster and harried across the face of Mount Rushmore. The audience has no idea why it's so important to stop Mason until a counterintelligence officer finally reveals what's at stake. "'Oh, just government secrets!' is the answer," Hitchcock explained in a famous interview with the filmmaker and critic François Truffaut. "Here, you see, the MacGuffin has been boiled down to its purest expression: nothing at all!"

Hitchcock may have popularized the term, but he certainly didn't invent the technique. Norris Lacy, a Penn State medievalist and former president of the International Arthurian Society, once wrote an article arguing that the Holy Grail itself was the ultimate MacGuffin, an object that launches the Knights of the Round Table on endless adventures but then mostly disappears from the narrative. Early Arthurian scholars often criticized the surviving stories for being full of "compositional flaws or incoherence," since plotlines seemed to appear then disappear with little warning or explanation. But these critics were missing the point, Lacy argues—or rather, missing the MacGuffins.

In the French Grail stories of the thirteenth century, and in many subsequent versions of the tale, it is revealed at the very beginning that it is Galahad's destiny to find the Grail. And yet all the other knights immediately vow to ride off in search of it, even though their search is doomed and they're not even clear what they're looking for. "If none but Galahad, in the French tradition, can succeed . . . why do others even expend the effort?" Lacy asks. Inzlicht would say it's because the effort itself is meaningful—and Lacy reaches a similar conclusion, reasoning that "the quest is, ironically, valorized by its very futility. It is the process, not the product, that counts."

In those interviews shortly before Joseph Campbell's death, when he talked about the high moments of his life, he wasn't remembering the feeling of crossing the finish line or the celebrations that followed. Those were MacGuffins. Instead, it was the moments when the outcome hung in the balance, when the struggle was fiercest, that he treasured: "the last eighty yards of the half mile," he said was when the peaks occurred. And I would go so far as to suggest that those moments would have been almost as meaningful even if he had lost the race—that the possibility of losing, in fact, was essential to the experience.

There are, at last count, nine different medals you can earn at the Comrades Marathon, the historic fifty-five-mile ultra that runs between the South African cities of Durban and Pietermaritzburg. While gold medals are awarded to the top ten men and women, the rest of the medals depend on hitting certain time standards. To earn a silver medal, for example, you have to break seven and a half hours; to earn a Robert Mtshali medal, named for the first Black runner to complete the race, you have to break ten hours; and to receive a Vic Clapham finisher's medal and be listed in the official results, you have to break twelve hours.

As each time threshold approaches, the stadium announcer and spectators count the seconds down. For the final twelve-hour deadline, a group of race marshals gathers in the finishing chute, and when the countdown reaches zero they lock arms to block the finish line. Either you make it or you don't. When I watched the race while reporting a magazine story in 2010, the final finisher, in 11:59:59, was a runner named Frikkie Botha, from nearby Mpumalanga, who placed 14,342nd. A stride behind, but not an official finisher after caroming off the race marshals' blockade, was forty-eight-year-old Dudley Mawona, from the historic inland town of Graaff-Reinet. Mawona accepted his fate with good grace. "I feel disappointed," he said. "But I am glad I was almost there." Both men resolved to return the following year.

The story I was reporting at Comrades was about the ultimate limits of endurance, and what struck me then was how finishing runners, no matter how tired they seemed, would accelerate as soon as the crowd began counting down—evidence, I figured, of the mind's role in determining physical limits. But my other lasting impression was of the stark delineation between success and failure, and the importance that runners and spectators alike attached to it. The woman next to me turned away rather than watch the final countdown. "I cried last year," she explained. "It's just too much to watch." When you line up at the start of Comrades, you know there's a very real chance that you won't finish, despite the months or years of training that you've put in. On that day in 2010, as is the case pretty much every year, more than a thousand runners who started the race didn't make it to the finish within twelve hours.

Failure is a delicate topic. A few months after returning from Comrades, I wrote an essay for the *Washington Post* that is, as far as I can recall, the only thing I've ever written that my mother told me she didn't like. There was controversy that year, as there often is, over the cutoff points in the Marine Corps Marathon. If you're not on fourteen-minute-mile pace by the time you hit the Fourteenth Street Bridge for the twenty-mile mark, for example, you're not permitted to continue. This is largely due to the logistical issues of shutting down traffic along major arteries in a big city, but it's seemingly at odds with the event's reputation as "The People's Marathon," emphasizing participation rather than competition. "I resent cutoff times," one *Post* reader complained prior to the race. "I think they're elitist ... and they discourage participation. If I pay to enter a race, I should not be kicked off the course just because I can't run fast."

A few years earlier, there had been a minor scandal when the coach of a large novice running group encouraged some of her athletes to take a shortcut in order to avoid getting pulled off the course at the Fourteenth Street Bridge. What sparked an uproar among purists was the coach's subsequent explanation: her charges were putting in 300 percent effort, she said. Many of the runners affected had indeed overcome remarkable odds to get as far as they did. Covering 22 miles, rather than 26.2, was still an amazing accomplishment. But it wasn't running a marathon. "One might contend that baseball would be a better game if four strikes were allowed instead of three," the philosopher John Rawls once wrote. It wouldn't be baseball, though.

What I argued, to my mother's discomfiture, was that the Dudley Mawonas of the world are the unsung heroes of events like Comrades and are essential to the warm glow of accomplishment felt by finishers even at events like Marine Corps, where competition is secondary. Success has no meaning without the possibility of failure, though the threshold that defines it will be different for everyone. Those who wish to qualify for the Boston Marathon have to hit standards that, for the fastest age groups, start at three hours. There used to be marathons in Japan where you were pulled off the course if you weren't on sub-2:30 pace at the halfway point. Conversely, there are people for whom the seven-hour cutoff at Marine Corps is out of reach, and who consequently should choose a race with a more forgiving time limit. The last-place finisher at the 2023 Honolulu Marathon, at 16 hours, 59 minutes, and 39 seconds, was thirty-six-year-old Andy Sloan. "To feel supported the whole way, even though I was the last person on the course, felt really, really good," he said afterward. "Knowing that I took the time to set a goal, and worked really hard to achieve it, it did mean a lot to me." It's not about the specific goal, in other words, but about having one that's not a foregone conclusion.

This is the bridge that connects challenging pursuits like running a marathon with the broader concept of exploring. An exploratory choice, in the explore-exploit literature, is one where you prioritize information over reward. If marathon legend Eliud Kipchoge wakes up one morning and decides he want to run 26.2 miles, there's no information to be gained. We know he can do it. If he decides he wants to run 26.2 miles *in under two hours*, that's a more interesting prospect. The reward of a successful completion is less likely, but he's exploring his limits and will learn something whether he succeeds or fails. If he aims for an hour and fifty minutes, on the other hand, the outcome—failure—is once again predetermined. The most satisfying challenge, in other words, is neither the hardest nor the easiest.

To study the science of "optimal challenge" you want a laboratory where you can precisely tune the degree of difficulty, where engagement is easy to measure, and where millions of people voluntarily show up to participate in your experiments for hours every day. That's the world of video game design, which has grappled with these ideas more or less since its inception. "Real fun comes from challenges that are always at the margin of our ability," Raph Koster wrote in his influential 2004 manifesto *A Theory of Fun for Game Design*. As you proceed through a game and gain skill, the gameplay becomes progressively more challenging—what Chris Crawford, a pioneering game designer with Atari in the early 1980s, called a "positive monotonic curve of results as a function of effort."

Typically, games allow you to choose various settings that adjust the challenge to suit your abilities and experience. In some cases, they use a technique called "dynamic difficulty adjustment," or DDA, to tweak the challenge automatically in response to how well you're doing. As early as 1981, according to Uppsala University game scholar Ernest Adams, an Intellivison game called *Astrosmash*—an imitator of the Atari classic *Asteroids*—would slow things down when you were almost out of lives. In casual racing games like *Mario Kart*, you may notice that the leaders get sideswiped by blue shells far more often than trailing racers, and the speed of racers is modulated by a "rubber band effect" that keeps everyone relatively close together. Disgruntled sports gamers actually filed a class action suit against Electronic Arts in 2020, claiming that games like *Madden NFL*, *FIFA*, and *NHL* deployed DDA to punish high-achieving players and nudge them to purchase performance-boosting loot boxes. (The suit was later dropped when Electronic Arts denied they used DDA in those particular games.)

The use of DDA is often explained in terms of Mihaly Csikszentmihalyi's theory of flow, which he said would occur "when the challenges are just balanced with the person's capacity to act." Other prominent theories of gaming emphasize psychological rewards like achievement and competence. But these accounts suggest that the perfect video game is one in which you always have a 50 percent chance of success—maximum uncertainty—since challenge and skill are perfectly balanced. That's hard to reconcile with the popularity of extremely easy "idle" games like *Cookie Clicker*, which require essentially no skill, and exceptionally difficult "Soulslike" games such as *Elden Ring*, where repeated failure is a precondition of progress.

In a 2022 paper, Imperial College game researcher Sebastian Deterding and several colleagues (including Marc Malmdorf Andersen) presented a theory of video games based on the goal of mastering uncertainty. Like Andersen's more general theory of play, their account draws on the principles of predictive processing, in which our actions are driven by the pleasure we get from minimizing the gap between what we expect and what we observe. Idle games, they point out, tend to increase rewards exponentially rather than linearly, so that even without any particular effort on your part you keep doing better than you expected. Soulslike games, in contrast, are so notoriously hard that even the slimmest morsels of progress exceed our expectations.

Deterding's grand theory of video gaming offers a useful template for thinking about the role of effort in the real world. The effort paradox and allure of optimal challenge don't mean that we need to turn life into a constant and never-ending struggle, dialing up the difficulty every time we're in danger of mastering something. Sometimes we might want the equivalent of a gentle idle game; other times, we're eager for the all-consuming struggle of a Soulslike game. Either way, exerting effort "seems to be the key route, maybe the only route, by which you can fulfill certain needs, like the needs for competence and mastery and maybe even self-understanding," Inzlicht told me. "You can't get those without pushing yourself." Campbell reached the same conclusion. His famous mantra, "follow your bliss," misled people into thinking that the hero's journey should be easy—a simple matter of doing whatever feels good. That's not what he believed, and not what his experiences as a world-class runner taught him. "I should have said follow your blisters," he later quipped.

But the motivational engineering that underlies modern video games also raises some uncomfortable questions. Computer games are just one element in a constellation of entertainment vehicles, from ancient storytellers to classic Hollywood films to the latest social media app. What they have in common is that they construct near-perfect prediction-error-reduction slopes—and their makers are getting dramatically better at it. "These apps, they deliver every time," Andersen told me. "They don't incentivize me to go and create my own slopes. I have no reason to go into the backyard and build a tree hut." What, then, does it mean to explore in a world where the urges that once drove us across oceans and into the unknown can be satiated, if only superficially, with the swipe of a finger?

12

THE FUTURE OF EXPLORING

For a brief instant before his canoe was sucked over the edge of a hitherto unknown twenty-five-foot waterfall in 2012, Adam Shoalts savored his triumph. For four years, he had been obsessed with the goal of paddling the entire length of the Again River, an obscure and almost unreachable waterway in the Hudson Bay Lowlands in northern Canada that, according to the annals of both historical and modern exploration, had never been traveled before. Now, as he and his battered boat plunged toward the frothing water and rocks below, he could finally confirm that existing maps of the area, derived from aerial and satellite imagery, really did omit some fairly significant geographical features.

Splash!

Shoalts grew up making birch bark canoes and roaming in the woods around his home in southeastern Ontario, a would-be explorer seemingly born in the wrong century. But he eventually spotted an opening. Sure, the whole world has been mapped—but how well? As late as 1916, the Geological Survey of Canada pegged the cumulative area of true blank spots on the country's maps as nine hundred thousand square miles, more than three times the size of Texas. The rise of aerial surveying soon filled in those gaps, but at the cost of boots-on-the-ground accuracy. Mapping landscapes from the air, Shoalts wrote in *Alone Against the North*, his 2015 account of the Again River expedition, "is no more like exploration than staring at the moon through a telescope in your backyard is akin to the Apollo moon landings."

He and his canoe survived the waterfall and made it to the end of the river. In the years since then, Shoalts has carved out a niche as "Canada's Indiana Jones" (as the *Toronto Star* described him) and an explorer-in-residence with the Royal Canadian Geographic Society, proclaiming to one and all that the Age of Exploration is not dead after all. His books are packed with adrenaline and hardship and breathtaking natural beauty, but it's his thirst for the unknown—the blank spots—that caught my attention. Finally, someone was expressing what I sought in my own far more modest wilderness trips, and articulating a vision of what true exploring could look like in the twenty-first century.

But then I read the reviews.

What is the future of exploration? As with any journey, it's easier to look back and see where we've come from than it is to figure out where we're headed and whether we're on the right path. Our instinctive thirst for the unknown helped us spread to every habitable corner of the globe, spurred us to unlock the mysteries of the natural world, and fueled the rise of our technological society. But where is it leading us now? Will the archetypal explorers of the future be establishing a colony on Mars? Mapping AI-generated worlds in virtual reality? Doggedly ticking off every entry on the *New York Times*'s annual list of top vacation spots to visit? Listlessly scrolling on their phones? Or can we find other ways—more broadly accessible, more productive, more meaningful—of scratching the itch?

Throughout this book, I've shifted back and forth between two distinct motivations for exploration. One is that it delivers useful rewards. Perhaps not immediately, and certainly not always, but on average, over the long term, we get scientific breakthroughs, new trade routes, better restaurant meals, and so on. The other motivation is that it feels good—again, not uniformly and not in the instantly gratifying way that, say, settling into a hot bath feels good, but in a way that delivers longer-lasting satisfaction and perhaps even meaning.

If I've blurred the distinction between these two motivations, it's because in the past they were always intertwined: fruitful forms of exploration evolved to feel good, so exploring in ways that felt good tended to be fruitful. That's no longer a safe assumption. As we gain the power to shape the world around us according to our whims and desires, we can no longer trust that what feels good is good for us. For most of our history, the sweet taste of sugar was a useful signal of easily available calories; now sugar is everywhere, and we've reluctantly realized that the fourth helping of dessert is probably a bad idea even if it still tastes good. Similarly, we've learned how to engineer hyperpalatable and instantly available forms of exploration that tickle our neural circuitry but don't lead us to anywhere useful.

"The primary way we're exploring today is in our tech," warns Mark Miller, the University of Toronto philosopher and cognitive scientist whose work on slope-chasing and slope-building we encountered in chapter 3. Social media, video games, and AI may offer us "the juiciest, shiniest, best slopes," he told me, but resolving their uncertainty doesn't teach us anything useful. "You're actually not learning that much about yourself, not learning that much about your environment. You're definitely not learning that much about other people, about how they actually work; or about the government, and how it actually works. You aren't learning about how to better live in this world—but you're getting all the sensations as if you're improving your grip in a much too easy way. That's dangerous. That sounds to me just like substance addiction." It's not enough to simply reduce uncertainty wherever we find (or create) it, in other words; we need to ensure that in doing so we're improving our ability to predict the world.

And there's another dichotomy to consider: between those explorations that reveal something new in an absolute sense and those that merely reveal something new to us. Part of what fascinated me about Adam Shoalts's exploits was that he had found a way of framing his adventures as genuinely new acts of discovery. His waterfall on the Again River wasn't just a nice view; it was one that, as far as he could ascertain, *no one in history had ever seen*. We'll come back to the question of whether that's really true. But as an aspirational goal, it's a bull's-eye, a perfect distillation of what I had always imagined it meant to explore. It requires some fairly major contortions to imagine that knowing about this waterfall will ever be useful to anyone. The Northwest Passage it's not. But the novelty of seeing it first justifies itself.

Not everyone agrees. The writer and explorer Kate Harris, in an article about the future of exploration, mocked the idea that Shoalts had "discovered" waterfalls—"when he accidentally canoed over them, no less"—in the long-established traditional territory of the Moose Cree First Nation. Another review called out his "misguided reverence for the lumbering spirit of European colonialism." More generally, scholars who study exploration have been rethinking their emphasis on what they've dubbed "firsting," the obsessive and often misleading focus on who did what first. In an editorial in Terrae Incognitae, the official journal of the Society for the History of Discoveries, historian Lauren Beck points out some of the inherent contradictions in firsting narratives: "How did Columbus become one of the world's greatest firsters through discovering a known land?" It's not just that the lands were inhabited, she points out; he was convinced he had reached Asia, a continent already known even to Europeans. And indeed, in his letters back to Europe, Columbus mostly used the verb *hallar*, meaning to observe or perceive one's location, rather than to discover it. The firsting narrative was only applied retroactively as rival European nations began to stake commercial and territorial claims in the New World.

My initial reaction to these criticisms was defensive—on both Shoalts's behalf and my own. After all, Shoalts had anticipated these objections and addressed them preemptively with two distinct arguments. The first was geographical: unlike more densely populated areas farther south, Canada's subarctic wilderness is both imponderably vast and all but uninhabitable. The Again River is located in the Hudson Bay Lowlands, a swampy, bigger-than-Minnesota wetland most notable for its polar bears and for having the highest concentration of bloodsucking insects in the world. Indigenous peoples certainty ventured into the area along its major rivers, but they considered it "sterile country," and there's little evidence of sustained precontact habitation. Given the exceptionally sparse population, along with the fact that Canada has something like three million lakes (no one has ever managed to count them properly) and innumerable rivers, creeks, and ponds, it's all but mathematically impossible that humans have visited every one of these waterways.

Of course, as Shoalts himself acknowledges, there's simply no way of knowing for sure whether anyone in previous centuries, let alone previous millennia, has ever visited a given place. But his second argument is that this doesn't matter, because exploration isn't just about hair-raising adventures but about "the generation of new geographical information that adds to humanity's stock of collective knowledge." If you've paddled the Again but you didn't file a report somewhere, preferably with an august geographical society brimming with cabinets of yellowing files that date back to earlier centuries, then you weren't engaged in the same task that he is.

This, then, is the case for firsting, in which the highest form of exploration is doing something that no one has done before, no matter how arbitrary the feat happens to be. It's the unspoken assumption that I've always carried with me in my own backcountry adventures: the more I can imagine myself as the first to venture into a trackless wilderness, the happier I am. If there's a path, Joseph Campbell said, it must be someone else's path, and therefore you're not on the adventure. But when you drag that assumption out into the daylight, it doesn't seem so convincing. Perhaps I've been chasing the wrong thing all this time.

The quintessential explorer, for me, has always been Étienne Brûlé, a Frenchman who came to North America as a teenager sometime around 1608 as a servant of Samuel de Champlain, the founder of New France. Unlike most explorers of the era, Brûlé left no record of his travels. It's not clear that he knew how to read or write, at least at first. But from Champlain and other contemporaries, we know that he was among the most prodigious firsters of his time. Most notably, he was the first European to see the Great Lakes: he likely made it to at least four of them, and it's possible he even saw the fifth, Lake Michigan, too. He is also thought to have explored what is now Pennsylvania, following the Susquehanna River all the way to its mouth on Chesapeake Bay.

My connection to Brûlé comes from having grown up on the banks of the Humber River, a crucial link in the ancient travel networks that Indigenous people blazed across the continent. The Humber route offers a shortcut between Lake Ontario and Lake Huron, skipping Lake Erie and saving days or even weeks on voyages to the interior of the continent. Brûlé traveled this way in 1615 on his exploration of the Great Lakes, and the linear parkland that runs alongside the river by my house in Toronto is called Étienne Brûlé Park. A few miles south, on a hillside overlooking the mouth of the river, there's a rough stone pillar marking the approximate spot where Brûlé first beheld Lake Ontario. The surrounding area is a wild oasis within the city; development was barred in the river valley after Hurricane Hazel flooded it in 1954, and as kids my friends and I spent long days playing elaborate games and constructing primitive shelters in what felt to us like a remote wilderness.

The Brûlé who shows up in Champlain's early journals is a man—a boy, really—of seemingly unquenchable curiosity. Champlain founded what is now Québec City in 1608; of the twenty-eight original settlers, Brûlé is thought to have been one of just eight to live through the first winter. To survive, the colonists needed help from the area's Indigenous inhabitants, and in 1610 Brûlé asked Champlain if he could go live among them. He wanted to "learn about their country, see the great lake, take note of the rivers and the peoples living along them; and discover any mines, along with the most curious things about those places and peoples, so that we might, upon his return, be informed truthfully about them," Champlain wrote. It was an audacious leap into the unknown, and Champlain was worried on Brûlé's behalf. "We asked him if it was his desire to go, for I did not wish to force him. But he answered the question at once by consenting to the journey with great pleasure."

There's something universal in that decision to plunge into the wilderness—to heed the Call to Adventure, as Joseph Campbell would say. But it also felt very personal to me, because Brûlé was plunging into *my* wilderness, the forests and hills and waterways where I now played. I was obsessed with trying to see the landscape with Brûlé's eyes, to imagine the panorama devoid of buildings and bridges, to feel the anticipation and uncertainty that must have accompanied each new bend in the river. I always imagined him alone. After all, he was exploring, and true exploring requires true wilderness in the sense of the word enshrined as a legal definition by the Wilderness Act of 1964: "an area where the earth and its community of life are untrammeled by man, where man himself is a visitor who does not remain."

That anthropocentric definition—wilderness is wherever we're not—has sparked plenty of debate in conservation circles about what types of places deserve protection, from whom, and for whose benefit. "If nature dies because we enter it," the environmental historian William Cronon pointed out in 1995, "then the only way to save nature is to kill ourselves." But the idea of wilderness as the absence of humans remains implicit in much of how we think and talk about exploring. Roderick Nash, the author of *Wilderness and the American Mind*, traces it back to the advent of herding, agriculture, and settlement ten thousand years ago, when lines both literal and metaphorical began to be etched into the land to delineate where human dominion started and stopped. And it still has a particular American resonance, thanks to the lingering influence of the Frontier Thesis, which attributes the emergence of a hardy and distinct national character in the 1800s to the century-long process of conquering the wild and untrammeled west.

To Nash, wilderness is an idea, or even a cultural invention, rather than a particular type of biogeographical place. To an extent, that means that perceptions become reality. "Regardless of what we might think about it today," he wrote, "Indians made the New World a greater, not a lesser, wilderness for the pioneer pastoralists." Traveling down the Humber River must have *felt* like a journey into the unknown for Brûlé. But objectively speaking, it clearly wasn't. That particular trip, in 1615, was a mission from Champlain and the Huron-Wendat people to ask for reinforcements from allies in the Susquehanna Valley for an upcoming battle with the Iroquois. The Huron-Wendat sent two canoes with "twelve of the most stalwart savages" to make the dangerous journey through Iroquois territory. Brûlé, once again, asked if he could tag along.

This is a pattern that recurs throughout the epic narratives of New World exploration. Alexander Mackenzie's voyages passed through terrain that was uncharted by Europeans, and his journals make for gripping reading—but you can't help but notice how little of his travel involved actually venturing into the unknown without any guidance. Instead, his progress amounted to a relay from tribe to tribe, shanghaiing locals into guiding his crew through each leg of the journey. "Thunder and rain prevailed during the night," he writes at one point, "and, in the course of it, our guide deserted; we therefore compelled another of these people, very much against his will, to supply the place of his fugitive countryman."

To historians and sociologists, of course, these sorts of revelations are old news. No one is putting up any new statues of Christopher Columbus these days. But if we're trying to plot the future of exploration, they present a riddle. The fact is that I still find the journeys of Brûlé and Mackenzie and others fascinating, even though I understand that their significance is different from what I used to imagine. I don't think it's just neocolonial nostalgia; the thrill of exploration really does persist even when others have preceded you. You can discern Brûlé's eagerness even through the thick mists of time and translation, even though he knew perfectly well that he was following in others' footsteps; and you can still seek and find echoes of the same feeling today. But if firsting isn't the highest aim of exploration, let alone a necessary precondition, then what replaces it?

I set up a choice above between absolute exploration, which uncovers something genuinely new about the world, and relative exploration, which merely reveals something new to you. Adam Shoalts seemed to be an avatar of absolute exploration in the thoroughly mapped modern world, but that turned out to be mostly an illusion—as it was for the vast majority of even the most famous explorers in history.

It's not that absolute exploration doesn't exist. There's a reason that tales of the early Everest expeditions or the first astronauts remain so compelling. But as far as our predictive brains are concerned, it's not the central point.

In his voyages and books since the Again River expedition, Shoalts has moved away from claims about discovering things. In 2017, he undertook a 2,500-mile canoe journey from west to east across Canada's Arctic, a challenge made all the more difficult by the fact that most rivers in the region run north-south. Much of the travel involved poling his canoe upstream or dragging it across the tundra. In 2022, he paddled more than 2,000 miles from his home in southern Ontario to the Arctic Ocean. In both cases, the central challenges were arbitrary. Rather than missions of discovery, they look more like Bernard Suits's "voluntary attempts to overcome unnecessary obstacles"—that is, like games.

Such game-playing, in Suits's view, is the highest ideal of human existence. He makes this argument by asking us to imagine a world in which all our material needs can be met at the press of a metaphorical button. No one wants for anything; no one needs to work for any instrumental end. What would we do with ourselves? How would we cope with the existential boredom? One option, Suits suggests, is to commit suicide. The other is to find something to do where overcoming difficulty is the intrinsic aim of the activity rather than the means to some other end. "Game playing," he writes, "makes it possible to retain enough effort in Utopia to make life worth living." You might argue that many of the things we do in our current world are both effortful and satisfying without being games. Building a house or cooking a meal can feel meaningful. But in Utopia, such things can be accomplished with a snap of the fingers, so to perform them manually is to voluntarily accept an unnecessary obstacle to their completion—to play a game, in other words. Suits calls this "occupational methadone," a transitional pastime for newcomers to Utopia.

Philosophers have been wrestling with Suits's claim ever since *The Grasshopper* was published in 1978. You don't need to take it literally in order to find it interesting. "Suits exaggerates when he calls game-playing the supreme good," the philosopher Thomas Hurka argues, "but it can still be one intrinsic good among many." In particular, according to Hurka, Suits's vision embodies a particularly modern view of the value of the journey rather than the destination. Classical philosophers like Aristotle argued that ends were inherently more valuable than the means used to attain them. Modern thinkers, on the other hand, generally see it the other way around, according to Hurka: Marx argued that workers would still see work as their "prime want" even after scarcity was overcome; Nietzsche argued that we're driven to exercise power just for the sake of doing so. "Marx and Nietzsche would never put it this way—their styles are far too earnest—but what each valued was in effect playing in games," he writes.

On this point, we can now see several different threads from earlier chapters begin to converge. To neuroscientists, the predictive processing model suggests that we're drawn not to a specific endpoint (the elimination of uncertainty) but to an ongoing process (the feeling of reducing uncertainty). To psychologists, the effort paradox ascribes meaning to the struggle itself, rather than its eventual rewards. To mythologists, the Holy Grail is just a MacGuffin, the pretext to be on a quest. And to philosophers—to some of them, at least—the voluntary attempt to overcome unnecessary obstacles is "the paradigm expression of modern values." Hurka, in an essay on *The Grasshopper*, cites the eighteenth-century German playwright and philosopher Gotthold Ephraim Lessing: "If God were to hold all Truth concealed in his right hand, and in his left only the steady and diligent drive for Truth, albeit with the proviso that I would always and forever err in the process, and to offer me the choice, I would with all humility take the left hand."

Shoalts, then, doesn't need an undiscovered waterfall to validate his voyages across Canada's north. The Again River too was just a MacGuffin. There is still plenty of exploring to be done even in a world where firsts are scarce. But this shift in perspective leaves us with another problem. If the journey itself is what matters, how do we know when we've arrived?

Étienne Brûlé's quest for reinforcements from the Susquehanna Valley didn't go well. He didn't show up for the battle, and Champlain and the Huron-Wendat were defeated by the Iroquois. Champlain didn't see him again for three years, at which point he arrived in Trois-Rivières with a group of Huron-Wendat warriors. When Champlain asked what had happened and where he'd been, Brûlé trotted out an impressive raft of excuses.

He had successfully summoned the warriors, he said, but they had arrived at the rendezvous two days late, after the battle was already lost. Stranded in hostile territory, he had decided to stick with his hosts in the Susquehanna Valley. He spent the next year living with them and exploring the surrounding regions, but while returning from his trip to Chesapeake Bay he and his traveling companions had been waylaid by an Iroquois ambush. After successfully escaping alone into the woods, he had discovered what it really meant to explore in the wilderness without friendly guides. He was unable to find his way or feed himself, and wandered aimlessly for several days. Finally, like Cody Sheehy, "he came upon a little footpath, which he determined to follow wherever it might lead, whether toward the enemy or not, preferring to expose himself to their hands trusting in God rather than to die alone and in this wretched manner."

Sure enough, he fell into Iroquois hands. After interrogating him, despite his denials that he was French, they began to torture him, tearing out his nails with their teeth, burning him with firebrands, and plucking his beard hair by hair. Then they noticed an Agnes Dei pendant around his neck; he warned them that touching it would incite the wrath of God, but one of them reached for it anyway. Fortunately for Brûlé, the sky happened to go dark at that moment, followed by "thunders and lightnings so violent and long continued that it was something strange and awful." This reversed Brûlé's fortunes, and he was then released and treated as an honored guest, invited to all sorts of "dances, banquets, [and] merry-makings." After living with the Iroquois for a while, he eventually promised to effect a reconciliation between the Iroquois and the French, and headed back to Huron-Wendat territory, where he once again lingered.

That, according to Brûlé, is why he was three years late. Subsequent historians have treated his tale, with its improbable twists and fortuitous thunderbolts, with a certain amount of skepticism. Champlain wouldn't have been impressed if he'd said "I couldn't get the warriors, so I decided to spend a few years exploring instead." We'll never know exactly what happened, but it's clear Brûlé wasn't in a huge rush to return to "civilization." And the first thing he asked of Champlain was for permission to return with the Huron-Wendat and continue exploring. Over the next decade, he pushed farther into the interior, possibly reaching as far as Duluth, at the western tip of Lake Superior.

Then things get complicated. In 1629, English privateers led by David Kirke and his four brothers, funded by the London-based Company of Adventurers to Canada, conquered Québec City. Accompanying the victors when they entered the city, according to Champlain, were four French traitors—one of whom was Brûlé, who claimed to have been taken by force. The rest of the French prisoners were sent back to France by the Kirkes, but Brûlé seemingly chose treason in order to remain in North America. He soon headed back into the interior, where according to the report of a French missionary he was killed and eaten by the Huron-Wendat in 1632 near the village of Toanché.

This grisly end only added to my fascination with Brûlé as a child. It seemed to validate what I imagined to be the perils of venturing into the untracked wilderness that had once surrounded my now-sedate home. Modern historians take this story too with a grain of salt: the missionaries disliked how Brûlé had adopted Indigenous customs, and may have thought that his sins deserved a suitably harsh storybook ending. The motive for his killing also remains unclear. One theory is that the Huron-Wendat feared that he was forging new trade links with other tribes, which would have eliminated their role as middlemen. But it's apparent that he was killed, and when the missionary Jean de Brébeuf arrived at Toanché in 1634 he found it deserted; a few years later, another missionary reported that the area was thought to be haunted by the ghost of Brûlé's sister, whose desire for revenge was blamed for the smallpox and other diseases then ravaging the Huron-Wendat population.

Brûlé is hardly the only famous explorer to venture into the wilderness, succeed beyond his wildest dreams, return to tell the tale... and then head back into the wilderness and keep repeating the pattern until his luck runs out. John Cabot was the first post-Viking European to reach North America; he died on his next voyage. Roald Amundsen was the first to traverse the Northwest Passage, in 1906, and the first to reach the South Pole, in 1911; after a long series of progressively more esoteric expeditions, he died in a plane crash in the Arctic in 1928. More common, perhaps, are trajectories like Edmund Hillary's: after climbing Mount Everest in 1953, he reached the South Pole in a converted tractor, climbed a handful of lesser Himalayan peaks, led an expedition searching for evidence of yetis, and flew on a ski-plane to the North Pole. He lived to the ripe old age of eighty-eight, but you get the sense that reaching the top of Everest, rather than sating him, left him even hungrier for more.

Bernard Suits called this "the Alexandrian condition," after Alexander the Great: "When there are no more worlds to conquer we are filled not with satisfaction but despair." As the name indicates, it's not a new phenomenon. But some observers think it's getting worse. Fred Previc, a psychologist at the University of Texas at San Antonio, argues that modern society is "hyperdopaminergic": goal driven, future oriented, and highly competitive in an environment of pervasive uncertainty, all of which fosters a heavy reliance on dopamine signaling in the brain. In a 2009 book called *The Dopaminergic Mind in Human Evolution and History*, he cites Alexander the Great, Christopher Columbus, Isaac Newton, Napoleon Bonaparte, and Albert Einstein as paradigmatic examples of the dopaminergic personality. They all achieved greatness; but they were also restless seekers whose insatiability eventually led to "consequences ranging from the merely embarrassing to the outright disastrous."

To Previc, the rise of dopaminergic societal values has led to spectacular successes: scientific discoveries, artistic creativity, "exploration to the farthest realms of the Earth and beyond," and more. But it has also come with costs that are no longer possible to ignore: not just endless wars, rising inequality, and declining leisure time, but existential threats to our environment and to our mental health. We can't (and wouldn't want to) simply return to the hunter-gatherer ways of our ancestors. "What can occur, however," he writes, "is the relinquishing of the dopaminergic imperative—the unquestioned sanctity of the human drive to explore, discover, acquire, and conquer/control." Some of this imperative is written into our genes, he acknowledges, but much of it is embedded in our culture.

Previc isn't alone in fighting the good fight against dopamine. A 2019 article in the *New York Times* shone a spotlight on the Silicon Valley minitrend of dopamine fasting, which involves abstaining from the steady stream of prediction error dished out by novelty-seeking modern activities like surfing social media, video games, and recreational drugs, not to mention more traditional triggers like eating and socializing. The idea is to weaken the link between these cues and the ensuing drop in prediction error, leaving more space to enjoy slower-paced pleasures like reading a book or walking in the woods. The image of tech bros studiously avoiding eye contact with each other produced a predictable wave of mockery, along with criticism of the oversimplified picture of dopamine's role in the brain. "The title's not to be taken literally," insisted the idea's progenitor, a San Francisco psychologist named Cameron Sepah. But the impulse behind dopamine fasting—the struggle to focus more acutely on what you have rather than being tormented by what you want—is an old, old story.

There's a postscript to Étienne Brûlé's saga. For several centuries after his death, his exploits were mostly forgotten, in part because the Frenchmen who wrote the histories of the era considered him a traitor and a sinner. Champlain edited out the account of his trip down the Humber River from later editions of his journals. In the eighteenth century, Brûlé was rediscovered by historians, and the stories they and their successors told about him reflected the ever-changing societal attitudes around them. He was an intrepid and insatiable explorer tragically murdered by savages; or he was an intrepid and insatiable explorer who got what he deserved. Those are the dueling narratives I grew up with.

In the 1970s, a historian named Lucien Campeau found evidence that, contrary to previous beliefs, Brûlé had actually returned to France from the wilds of Canada twice before his death: once from 1622 to 1623, and then again from 1626 to 1628. In 2010, another historian, Éric Brossard, unearthed more documents from archives in Paris and Champigny-sur-Marne, Brûlé's birthplace, that fleshed out the details of his trips back. Brûlé, it turns out, wasn't an incorrigible *coureur de bois* who preferred death on the frontier to life in the city. Thanks to the money he made fur-trading overseas, he returned to Paris as a prosperous merchant. He shows up in baptismal registers as a godfather to several children; he loans money and makes investments, entrusting their management to a notary. He buys two houses, one in Paris and one in Champigny. He marries a woman named Alizon Coiffier. To all appearances, Brossard believes, he was preparing to settle down to a comfortable life in France, one that after years of hard work was finally within his grasp—after one more trip to the frontier.

It was on the voyage back to Canada in 1628 that his ship was seized by the English, who took him back to London as a prisoner, setting in motion a sequence of events that ended with him being clubbed to death in Toanché, four thousand miles from his home and family. It changes how you think about Brûlé and his voyages. We can't know what was on his mind, but it's not hard to imagine the tug-of-war between adventure and safety; between having and wanting. In one form or another, we've all been there.

There's one final dichotomy that I've been wrestling with ever since I began thinking about exploration. My interest in the topic arose in part from the desire, as I stared down the tracks at the approaching headlights of middle age, to keep exploring. I wanted to understand what drove me, and push back against what I now understand to be a natural and mathematical arc of declining exploration with the passage of time. I figured the subtitle of this book might be something along the lines of "The Case for Embracing the Unknown." But I was also nagged by the suspicion that it's just as important to understand when to *stop* exploring, or at least when to pause for a while and appreciate the place you've arrived at. Relinquishing the dopaminergic imperative might help keep you out of the cooking pot; it might also be the path to a happier and more sustainable society.

I haven't found any universal formula for choosing between these two paths. Sometimes you need to explore; sometimes you need to exploit; and sometimes, as Kenny Rogers suggests, you need to walk away. There are times too when you can do both at once: appreciate where you are while simultaneously gazing at the horizon. Nothing has cultivated this spirit of nondualism in me as effectively as my years as a competitive runner. So many of my wildest dreams came true—winning races, hitting times, qualifying for national teams—that I soon came to understand that achieving one goal immediately gives birth to the next one. There's always another rung on the ladder, another land to be discovered. If you manage to stop caring about the next rung, the whole pursuit loses its meaning. But if you imagine that lasting happiness awaits there, you're doomed to disappointment. The lessons are even sharper now that I'm in my late forties and getting slower with every passing day. You can't tie your self-worth to the prospect that you'll be better tomorrow, but probing the outer limits of my capabilities remains among the most profound and meaningful forms of exploration in my life. "We're not given knowledge about our true skill level," Charley Wu told me. "We have to acquire that knowledge through interrogating ourselves, through testing ourselves."

So, instead of advice on how to explore more or less, I'll offer instead five rules for how to explore *better*, gleaned from the research described in the preceding chapters. They're not iron-clad rules, of course. They're more like guiding principles to keep in mind, amid the complexity of real-world decision-making, when you find yourself on the horns of an explore-exploit dilemma.

1. EXPLORE *THEN* EXPLOIT

This is a rule that arises naturally from the mathematics of multi-armed bandits with different time horizons. But it's also a pattern that has been field-tested in Dashun Wang's career-trajectory studies. To exploit without first having explored is to potentially miss the best opportunities; but to explore indefinitely is to be a dilettante. It's the combination that kills.

2. SEEK THE UNCERTAINTY SWEET SPOT

The well-traveled highway isn't particularly interesting, but neither is the impenetrable jungle devoid of landmarks and filled with man-eating tigers. To get the most out of exploring, you have to find intermediate levels of uncertainty, which offer the richest opportunities to resolve uncertainty.

3. PLAY MORE

To find that uncertainty sweet spot, trust your instincts. Curiosity, interest, and just plain having fun are the cognitive manifestations of your brain happily reducing prediction errors. The hardest part, for adults living in a complicated and fast-moving world, is creating the space needed and giving yourself permission to sometimes follow your nose.

4. MINIMIZE REGRET

Sometimes you make a bold and adventurous choice when ordering at a restaurant, and the resulting meal turns out to be terrible. That doesn't mean you were wrong to explore; regret is an unavoidable part of being human. But the best way to minimize regret, according to a half-century of decision science research, is to choose optimistically. The "Upper Confidence Bound" approach suggests following the path with the most favorable best-case outcome.

5. EMBRACE THE STRUGGLE

We have many sources of novelty at our disposal in the modern world. We can travel the globe or start a business; we can read a novel or surf the internet. There's room in our lives for all of these ways of tickling the exploring circuitry in our brains. But bear in mind that the most challenging paths often turn out to feel the most meaningful—not in spite of the effort required, but because of it.

Each of us, of course, will weigh the pros and cons of a given set of choices differently. It's clear that some people value novelty more highly than others, perhaps because of their dopamine receptor genes or other quirks of biology, perhaps because of the vagaries of their life history. But it's also clear that these inclinations are not fixed in stone. They can vary across the lifespan, or from moment to moment. The predictive processing theory of the brain suggests that our inclination to explore rather than exploit is influenced by the balance between bottom-up sensory information collected from the environment and top-down predictions formulated based on prior experiences and expectations. And it suggests ways of tipping that balance in one direction or the other.

Most generally, as Moshe Bar's work on "overarching states of mind" suggests, we're primed to explore when we're thinking broadly rather than narrowly. Stress and anxiety cause you to narrow your focus, zooming in on perceived threats both literally (you're less aware of peripheral visual information) and conceptually (you have a harder time coming up with creative ideas). Reducing your stress levels puts you in a more exploratory frame of mind. Deliberately thinking about the bigger picture can accomplish the same thing: instead of focusing on the narrow and immediate consequences of a decision, think about its broader long-term implications.

Exercise, too, seems to leave us in a more expansive and exploratory frame of mind. And there's even a study—in mice—that found links between exploratory behavior and the microbes in their guts. A microbiome transplant from exploratory mice turned stay-at-home mice into adventurers, and vice versa. Maybe someday, instead of trying to encourage ourselves (or trick ourselves) to think broadly, we'll be able to simply take a probiotic exploration pill.

But the study that sticks in my mind most stubbornly is simpler and less fanciful. It's from a group of researchers at University of California, Irvine, including Chuansheng Chen, who led the first study of the DRD4 dopamine receptor and human migration. The scientists ran genetic tests on 311 of the "oldest-old" residents of a retirement community called Leisure World, in Laguna Woods, California, ranging in age from 90 to 109. Relative to a comparable group of younger people, these super-agers were 62 percent more likely to have the "explorer's gene" version of the DRD4 dopamine receptor. Similarly, mice who were bred to have nonfunctional DRD4 receptors were less exploratory and didn't live as long as regular mice, even if they were reared in enriched environments that offered plenty of opportunity for physical activity and mental stimulation.

That's not enough evidence to declare that exploring makes you live longer. Maybe the DRD4 gene simply prods people to get more exercise; maybe it helps you absorb life's unexpected slings and arrows with less health-harming stress. The effect was also stronger in women than men: the researchers hypothesize that the longevity benefits of exploring might be partly counteracted by its association earlier in life with risky (and male-biased) behaviors such as drug abuse. There's no free lunch. Still, the findings conveniently overlap with the conclusion of all those other lines of evidence—the evolutionary biology, the neuroscience, the computational cognitive science, the psychology—and with my deeply held but evidence-free gut instinct: on average, and in the long run, exploring pays off.

EPILOGUE

In early July 2024, Lauren and I were back in the mountains with our long-suffering kids. We'd flown to Europe for my brother-in-law's wedding, then tacked on a six-day hut-to-hut hiking trip high in the Spanish Pyrenees, along a circular route known as the Carros de Foc, or Chariots of Fire. The trip had many charms—vivid, almost fluorescently blue alpine lakes; towering granite rock faces still draped in snow; hearty multicourse dinners at cheerful communal dinner tables—which seemed all the more charming in contrast to the three hectic days we spent in Barcelona after the hike. Protestors there were spraying tourists with water and marching with TOURISTS GO HOME! signs. As we shuffled uncomfortably along La Rambla amid the overwhelming crush of our fellow seekers, the protests didn't seem unreasonable.

Back in 1773, while Captain Cook was island-hopping in Polynesia and the age of European discovery was in full swing, the French philosopher Denis Diderot uncorked a scathing critique of "the active, difficult, wandering, dissipated life of an explorer." It wasn't just the rapaciousness of the colonial enterprise that he decried, though he was no fan of the unending search for "continents to invade, islands to ravage, peoples to despoil, subjugate, and massacre." He also thought that the explorers themselves were degraded by their own wanderlust, spurred on by a contemptible mix of "ambition, misery, curiosity, I know not what restlessness of spirit, the desire to know and the desire to see, boredom, the dislike of familiar pleasures." There was, he wrote, "no state more immoral than that of the continual traveler."

Diderot was a little over the top, at least in my view, but when I started working on this book I had my own misgivings about the exploratory urges that always seemed to be nipping at my heels. Were they a productive force, leading me onward to fulfilling adventures and productive new horizons? Or were they simply luring me into a never-ending cycle of frivolous consumption and unfulfilled desire? After digging deep into the workings of our reward circuitry and the imperatives of a predictive brain, I'm now confident that the answer is a hearty "it depends."

I've become a convert to what Paul Bloom calls motivational pluralism: we want many different things, and the same action might be motivated by a different mix of factors at different times. Sometimes we explore with the goal of maximizing some specific outcome, whether it's corporate profit or marital bliss, framing the decision as an explore-exploit choice. Other times we're simply trying to follow our curiosity and learn about the world, tapping into a different and more playful set of cognitive pathways. Sometimes we explore reflexively, without conscious awareness of why we're doing it; this is the predictive circuitry that casinos and social media algorithms coopt. Other times we explore deliberately, mindful of both the potential benefits and the opportunity costs to ourselves and others.

In the Pyrenees, I tried my best to explore mindfully. The location itself was a deliberate choice, a region that isn't yet on the standard international tourist circuit but is nonetheless relatively accessible. I didn't need a trackless wilderness, or Parks Canada warning me to stay away, to make it feel worthwhile. During our six days of hiking, we didn't encounter a single native English speaker. Virtually everyone else on the trails was from Spain, with only a few exceptions: a father and son from Denmark, an older couple from France. Navigating the elaborate social customs of the huts, where as many as sixty people might share bunks in a single room and queue for a single shower, was a greater source of uncertainty to us than getting over nine-thousand-foot passes.

I had the route's GPS waypoints downloaded onto my phone, just as I had in Newfoundland two years earlier. This time, though, I made a concerted effort to keep my phone in my pocket and rely instead on a paper map. Doing so forced me to look around and to understand where I was relative to where I wanted to go. Instead of checking my phone to confirm that we wanted to stay left at the next fork, I had to read the landscape and compare it to the contour lines on my map: aim for the col to the left of that glacier on the far side of the river. The benefit was that now I really *saw* the col, and the glacier, and the river. And my approach was reminiscent of the hippocampus-preserving advice that Véronique Bohbot gives about how to use GPS in the city: start with a bird's-eye view of the route, figure out where you're going... and then close the app until you need it.

To be sure, our voyage in the Pyrenees felt more like Nellie Bly's than, say, Alexander Mackenzie's: rather than blazing a new route, we were simply trying to complete a well-worn circuit in the time allotted. But I was trying to focus less on endpoints. The novelist Cormac McCarthy, when he read a novel, used to start by flipping to the end and reading the last page. "It's because he was not interested in the uncertainty reduction that came with knowing the outcome of plot," McCarthy's friend and former colleague at the Santa Fe Institute, David Krakauer, told *Scientific American*'s Christie Aschwanden. Instead, he was interested in a different form of uncertainty: how a writer thinks and how they encode reality. Simply knowing how the story ends, Krakauer explained, is like knowing the final score of a basketball game without having watched any of the action. The score matters in a MacGuffin-like way, but what defines the experience is how you get there and what adventures you encounter along the way.

I was trying to follow a similar dictum. My primary goal, I kept reminding myself, wasn't to make it to the next hut; it was to be in the mountains, journeying through an unfamiliar landscape toward a destination shrouded in uncertainty. "We actually enjoy, cognitively, not knowing," Krakauer explained, "and we pay huge sums of money to not know things." This less goal-oriented mindset doesn't come naturally to me, I'll admit. But it's true that we had paid a lot of money to be in the Pyrenees, and I did my best to embrace it. I let the kids wander off the trail more frequently than I usually do—and followed them off it to explore a cave that turned out to have rudimentary stone furniture and a fireplace in it, and to scramble up a slope to a patch of unmelted snow where we cooled off with a snowball fight. Arriving at the hut at the end of each day was still a magical feeling, but I now understood more clearly that it wasn't the reason we were out there.

In February 2024, a case of unopened O-Pee-Chee hockey cards from 1980 sold at auction for $3.72 million dollars, a new record for the most expensive lot of sealed sports cards. A family in Regina, in the Canadian province of Saskatchewan, had unexpectedly rediscovered the case in their basement and realized that it dated from Wayne Gretzky's rookie year. Among the thousands of cards in the case, the odds suggest that there are twenty-five to twenty-seven pack-fresh Gretzky rookie cards. A single Gretzky rookie in mint condition has sold for as high as $3.75 million, which makes the auction price seem like an amazing bargain. But cards in 1980 were still cut with wires that can produce ragged edges, were often off-center, and sometimes even had a piece of gum stuck to them, so it's far from guaranteed that the case contains even a single truly mint Gretzky.

What I found fascinating in the news coverage about the auction was that all the experts seemed to agree that the case was unlikely to be opened—perhaps ever. For now, no one knows how many Gretzky cards it contains and how good their condition is, and the case's valuation reflects that uncertainty bonus. But once it's open, you can't dream about what might be in there; you have to accept the reality of what *is* there. "The allure of the great potential of what 'could' be inside the case is what gives the item its enhanced value," Michael DiStefano of *Hockey News* explained. "So as long as the case remains sealed, it should not only hold value but appreciate even further over time." The mystery-shrouded hockey cards reminded me of Schrödinger's cat, suspended in a state of simultaneous life and death until someone opens its box. They also brought to mind a line from the American environmental writer Aldo Leopold: "Of what avail are forty freedoms," he asked, "without a blank spot on the map?"

I felt the truth of Leopold's rhetorical question in my bones when I first read his book *A Sand County Almanac*, and it filled me with panic. The blank spots, after all, are disappearing at an astounding pace. For a long time, much of my travel was devoted to rushing around the globe trying to catch some of the remaining blank spots before they too were fully mapped and commoditized. That sense of urgency was particularly pronounced during a 2012 canoe trip down the remote Snake River, in the Yukon. The territorial government had recently overturned the recommendations of a planning commission that the watershed be preserved, and a vast swath of roadless wilderness was in imminent danger of being thrown open to mining. Around the campfire, my friends and I debated the balance between preserving wilderness for the seemingly exclusive benefit of a few privileged carpetbaggers from down south like ourselves versus letting people in the Yukon exploit their resources however they wished. The Snake was worth safeguarding, I argued, even if I was no longer allowed to visit it. Simply knowing that such places exist has immense value, far beyond anything we can extract from it—like leaving a case of hockey cards unopened, perhaps forever.

My fears about the imminent end of exploration were based on the assumption that blank spots on the map are a finite resource that might someday run out. But I've come to appreciate a much broader conception of what counts as a blank spot worth exploring: not just swaths of untrammeled wilderness, or even packs of unopened hockey cards, but obscure corners of my neighborhood, and unexpected career opportunities, and unfamiliar cuisines. The trajectory of adulthood is toward ever greater efficiency, narrower focus, and well-worn routines that make each day more and more similar to the last. Exploration is the antihabit, the antidote to a diminished palette of life choices. We're wired to seek out the unknown, to embrace the challenges we find there, and to find meaning in the pursuit.

I recently bought some inexpensive kayaks that I can carry a few hundred yards down the hill from my house in the heart of Toronto to the banks of the Humber River—the same river that Étienne Brûlé traveled down four hundred years ago, and that Indigenous people paddled for centuries before that. During the pandemic, I spent a lot of time with my kids in the patchwork of parkland and untamed forest that runs alongside the river. We foraged for thimbleberries and wild grapes and mustard greens, and caught glimpses of beavers and coyotes and deer. As the kids got older, I began to let them wander farther afield on their own. I watched them building forts and discovering secret routes, and saw patterns that are imprinted in all of us.

The kayaks have given us yet another perspective on this strip of urban wilderness. From the water, the paved bike path and the areas of mowed grass are invisible; the roads, other than the occasional bridge, are just a distant hum lost in the burbling of the rapids. Toronto is a city of three million people, but from the water it's surprisingly easy to imagine what it must have been like centuries ago. After a rainstorm a few weeks ago, we rode the swollen currents downstream and exited the main channel to explore a marsh that's usually inaccessible at lower water levels. I've been playing along this river since I was a kid, and know most of its hidden corners and byways, but the marsh isn't reachable by land. We paddled past egrets and cormorants, and startled a great blue heron that flew out of the reeds a few yards ahead of us. At the far end of the marsh, we came to a fallen tree, long dead, with more than twenty turtles basking in the sun on its time-smoothed trunk and branches. Some were as small as a golf ball, others as big as a serving platter. We drifted quietly for a while, keeping our distance from the turtles, taking in the new sights and sounds.

I still love traveling to far-flung and unfamiliar places, and I still worry about the disappearing blank spots on the physical map, as development and exploitation and environmental degradation roll inexorably onward. I still struggle to strike the right balance between seeking novelty and sticking with sure bets in my daily life, between liking and wanting, between pursuing goals and making space for digressions and diversions. But I'm less worried about running out of opportunities to explore—because they're all around us.

ACKNOWLEDGMENTS

Writing this book was—not to wear the metaphor too thin—a voyage into the unknown. I immersed myself in several new-to-me areas of science, which involved leaning even more heavily than usual on the generosity of researchers who shared their time, insights, and expertise. I'm enormously grateful to all of them, in particular those who helped read and fact-check successive drafts. All remaining errors are, of course, solely my responsibility.

Matt Harper, my editor at Mariner Books, turned out to be exactly the guide I needed for the writing process: when things weren't going well, we could at least talk about running. I'm also grateful to Peter Hubbard for believing in me, and to the rest of the team at Mariner, including Ivy Givens, Tavia Kowalchuk, and Lindsey Kennedy. My agent, Rick Broadhead, has been both a trusted friend and a dogged advocate for a decade and a half; I appreciate both roles.

I've accumulated innumerable other debts of gratitude over the past several years. Rodger Kram read an early draft and spurred me to think carefully about language use. The Bentley Historical Library at the University of Michigan located some of Merrill Flood's correspondence. The IEEE History Center granted permission to quote from its Engineering and Technology History Wiki oral history interview with Frédéric Kaplan. Julie Witmer patiently waded through my garbled sketches and contradictory notes to produce beautiful maps.

I'm also grateful to the many readers who have engaged with the ideas I've written about over the years. This engagement is by far the most satisfying part of my job, and the feedback and suggestions—

and yes, sometimes the critiques—play a crucial role in shaping what I write.

Most of all, I'm grateful to my family: to my parents, Moira and Roger, for giving me so many opportunities to explore; to my wife, Lauren, for exploring with me; and to my children, Ella and Natalie, for their curiosity, joy, and willingness to endure ten hours in the mud.

NOTES

INTRODUCTION

2 *"98 percent of visitors"*: Bill Bryson, *Notes from a Big Country* (London: Doubleday, 1998).

3 *"less risky adventures"*: Parks Canada's website for the Long Range Traverse was revised in 2024 and no longer warns visitors away.

4 *"It was intoxicating"*: "Far, Maybe Too Far, into the Yukon," *New York Times*, August 23, 2012.

4 *"our preciously rationed vacation days"*: "Mud, Leeches and Stunning Beauty in Tasmania," *New York Times*, January 25, 2013.

5 *a book called Endure*: *Endure: Mind, Body, and the Curiously Elastic Limits of Human Performance* (New York: William Morrow, 2018).

5 *the New York Times bestseller list:* It appeared on the obscure (and since discontinued) Sports and Fitness list, peaking at number three in March 2018.

6 *a lab affiliated with the University of Maryland:* The Laboratory for Physical Sciences, in College Park, Maryland, was founded in 1956 "to drive physical sciences research in future information technologies to prevent technological surprise."

7 *"Exploration and Exploitation in Organizational Learning"*: *Organizational Science* 2, no. 1 (1991).

7 *should have shared Simon's Nobel:* "James G. March, Professor of Business, Education, and Humanities, Dies at 90," Stanford Graduate School of Business, October 29, 2018.

8 *45 percent of our actions:* Wendy Wood and David Neal, "The Habitual Consumer," *Journal of Consumer Psychology* 19, no. 4 (2009).

9 *Even regular commuters:* Shaun Larcom et al., "The Benefits of Forced Experimentation: Striking Evidence from the London Underground Network," *The Quarterly Journal of Economics* 132, no. 4 (2017).

9 *a progressively bigger dose:* To be clear, this is a major oversimplification of the complexities of drug addiction, which also involves changes in how brain cells respond to drugs.

10 *The Latin word explorare*: "Explore," *Oxford English Dictionary* (Oxford: Oxford University Press, 2016).

10 *"to scout the hunting area"*: Michiel de Vaan, *Etymological Dictionary of Latin* (Leiden: Brill, 2008).

11 *"the great suburban Everest"*: John Bryant, *Chris Brasher: The Man Who Made the London Marathon* (London: Aurum Press, 2012).

11 *a Norwegian explorer and adventurer named Helge Ingstad*: For details of the rediscovery of Vinland, see Benedicte Ingstad, *A Grand Adventure: The Lives of Helge and Anne Stine Ingstad and Their Discovery of a Viking Settlement in North America* (Montreal & Kingston: McGill-Queen's University Press, 2017); and Birgitta Wallace, "The Discovery of Vinland," in *The Viking World*, ed. Stefan Brink (London: Routledge, 2008).

12 *a young boy named Clayton Colbourne*: Mary MacKay, "Clayton Colbourne Is on His Home Turf," *Prince Edward Island News*, July 23, 2014.

CHAPTER 1: THE GREAT HUMAN EXPANSION

17 *Mau the navigator predicted:* This account of the voyage of the Hokule'a is based primarily on Ben Finney, *Hokule'a: The Way to Tahiti* (New York: Dodd, Mead & Company, 1979); and Sam Low, *Hawaiki Rising: Hokule'a, Nainoa Thompson, and the Hawaiian Renaissance* (Honolulu: University of Hawai'i Press, 2013).

18 *the mathematical laws of diffusion*: James S. Clark, "Why Trees Migrate So Fast: Confronting Theory with Dispersal Biology and the Paleorecord," *American Naturalist* 152, no. 2 (1998).

18 *the same equations to model how human populations*: Joaquim Fort, "Biased Dispersal Can Explain Fast Human Range Expansions," *Scientific Reports* 10 (2020).

18 *"perpetual Viceroy and Governor"*: Christopher Columbus, Journal of Christopher Columbus (During His First Voyage, 1492–93), and Documents Relating to the Voyages of John Cabot and Gaspar Corte Real (London: Hakluyt, 1893).

19 *the psychologist Daniel Berlyne*: D. E. Berlyne, *Conflict, Arousal, and Curiosity* (New York: McGraw-Hill, 1960).

19 *"Because it's there!"*: "Climbing Mount Everest Is Work for Supermen," *New York Times*, March 18, 1923.

19 *"Exploration is in our nature"*: Carl Sagan, *Cosmos* (New York: Random House, 1980).

19 *according to J. S. Johnson-Schwartz*: J. S. J. Schwartz, "Myth-Free Space Advocacy Part I – The Myth of Innate Exploratory and Migratory Urges," *Acta Astronautica* 137 (2017).

20 *according to the historian Felipe Fernández-Armesto*: Felipe Fernández-Armesto, *Pathfinders: A Global History of Exploration* (New York: W. W. Norton, 2006).

20 *starting in earnest about fifty thousand years ago*: Nailing down the dates of ancient human migrations is a tricky and ongoing process. In *Neanderthal Man* (New York: Basic Books, 2014), Svante Pääbo pegs the appearance of fully modern humans in Africa at fifty thousand years ago, after which this "replacement crowd," as he calls them, rapidly spread to "almost every habitable speck of land on the planet." In "The Great Human Expansion," *PNAS* 109, no. 44 (2012), Brenna Henn et al. give a broader range of sixty thousand to forty-five thou-

NOTES

sand years ago for this rapid dispersal. In *Who We Are and How We Got Here* (New York: Pantheon Books, 2018), David Reich suggests that modern humans spread from Africa and the Near East less than fifty thousand years ago.

20 *two million years ago:* Pääbo notes *H. erectus* fossils in Georgia dated to 1.9 million years ago.

20 *as early as 270,000 years ago:* Cosimo Posth et al., "Deeply Divergent Archaic Mitochondrial Genome Provides Lower Time Boundary for African Gene Flow into Neanderthals," *Nature Communications* 8 (2017).

20 *This concept of behavioral modernity:* Richard Klein, "Anatomy, Behavior, and Modern Human Origins," *Journal of World Prehistory* 9, no. 2 (1995), lays out the case that behavioral modernity emerged quite suddenly between fifty thousand and forty thousand years ago. Sally McBrearty and Alison Brooks, "The Revolution That Wasn't: A New Interpretation of the Origin of Modern Human Behavior," *Journal of Human Evolution* 39, no. 5 (2000) argue for a more gradual emergence starting much earlier.

21 *twenty thousand miles away:* Luke Matthews and Paul Butler, "Novelty-Seeking DRD4 Polymorphisms Are Associated with Human Migration Distance Out-of-Africa after Controlling for Neutral Population Gene Structure," *American Journal of Physical Anthropology* 145 (2011), estimate the total migration distance from Africa for the Surui, in Brazil, to be 32,124 kilometers.

21 *about eight hundred years ago:* Atholl Anderson, "The Chronology of Colonization in New Zealand," *Antiquity* 65, no. 249 (1991); but as Christina Thompson notes in *Sea People* (New York: HarperCollins, 2019), some estimates now push the date even closer to the present.

21 *in a National Geographic article:* David Dobbs, "Restless Genes," *National Geographic* 223, no. 1 (2013).

21 *"I don't like this book":* quoted in Sam Low, *Hawaiki Rising: Hokule'a, Nainoa Thompson, and the Hawaiian Renaissance* (Honolulu: University of Hawai'i Press, 2013).

21 *Ancient Voyagers in the Pacific:* Andrew Sharp, *Ancient Voyagers in the Pacific* (Harmondsworth: Penguin, 1957). The book was initially published by the Polynesian Society in 1956.

22 *The Spanish navigator Álvaro de Mendaña:* For a good overview of the chronology of Polynesia exploration, see Thompson, *Sea People*.

22 *"people without skill":* quoted in Ben Finney, *Voyage of Rediscovery: A Cultural Odyssey Through Polynesia* (Berkeley: University of California Press, 1994).

22 *"descendants of Adam":* Andrew Sharp, ed., *The Journal of Jacob Roggeveen* (Oxford: Clarendon, 1970).

22 *"of which volcanic shocks":* Crozet's *Voyage to Tasmania, New Zealand, the Ladrone Islands, and the Philippines in the Years 1771-1772* (London: Truslove & Shirley, 1891).

23 *his mother's youngest sister:* Eleanor Harmon Davis, *Abraham Fornander: A Biography* (Honolulu: University Press of Hawaii, 1979). Details of this supposed affair are highly speculative.

NOTES

23 *in the foothills of the Hindu Kush*: Fornander's Polynesian theories are summarized in Thompson, *Sea People*.

24 *"Vikings of the Pacific"*: Peter Buck (also known as Te Rangi Hiroa), *Vikings of the Pacific* (Chicago: University of Chicago Press, 1959). The title of the original 1938 edition was *Vikings of the Sunrise*.

24 *"very long voyages"*: *The Endeavour Journal of Sir Joseph Banks, 1768-1771* (Sydney: Angus & Robertson, 1962).

24 *eaten by cannibals*: Briony Leyland, "Island Holds Reconciliation over Cannibalism," *BBC News*, December 7, 2009.

24 *"an object of ambition"*: John Williams, *A Narrative of Missionary Enterprises in the South Sea Islands* (London: John Snow, 1837).

24 *"cosmopolites by natural feeling"*: Horatio Hale, *Ethnography and Philology* (Philadelphia: C. Sherman, 1846).

24 *"fired by the lust"*: Raymond Firth, *We the Tikopia: A Sociological Study of Kinship in Primitive Polynesia* (London: George Allen & Unwin, 1936).

24 *scraps of wood*: Finney, *Voyage of Rediscovery*.

25 *"One of the most provocative"*: K. R. Howe, "Voyagers and Navigators: The Sharp-Lewis Debate," in *Texts and Contexts: Reflections in Pacific Islands Historiography* (Honolulu: University of Hawaii Press, 2005).

25 *"I felt that Andrew Sharp"*: Pei Te Hurinui Jones, "A Maori Comment on Andrew Sharp's 'Ancient Voyagers in the Pacific,'" *Journal of the Polynesian Society* 66, no. 1 (1957).

25 *seventy-four islands within a four-thousand-mile radius*: Lars Eckstein and Anja Schwartz, "The Making of Tupaia's Map: A Story of the Extent and Mastery of Polynesian Navigation, Competing Systems of Wayfinding on James Cook's Endeavour, and the Invention of an Ingenious Cartographic System," *Journal of Pacific History* 54, no. 1 (2019).

25 *"same Nation"*: *The Three Voyages of Captain Cook Round the World* (London: Longman, 1821).

26 *"prove to any perceptive mind"*: Andrew Sharp, "Polynesian Navigation to Distant Island," *Journal of the Polynesian Society* 70, no.2 (1961).

26 *a hundred thousand possible drift voyages*: R. Gerard Ward et al., "The Settlement of the Polynesian Outliers: A Computer Simulation," *Journal of the Polynesian Society* 82, no. 4 (1973); see also Thompson, *Sea People*.

27 *"while flying before the trades"*: Finney, *Hokule'a*.

28 *would take a million years*: Clement Reid, *The Origin of the British Flora* (London: Dulau, 1899).

28 *surprisingly long distances*: Clark, "Why Trees Migrate So Fast."

29 *places like Crete and the Indonesian island of Flores*: Dylan Gaffney, "Pleistocene Water Crossings and Adaptive Flexibility Within the Homo Genus," *Journal of Archaeological Research* 29, no. 2 (2021).

29 *one view is that the water crossings*: Iain Davidson and William Noble, "Why the First Colonisation of the Australian Region Is the Earliest Evidence of Modern Human Behaviour," *Archaeology in Oceania* 27, no. 3 (1992).

30 *"It is a complex enough task":* Thomas Leppard, "The Evolution of Modern Behaviour and Its Implications for Maritime Dispersal During the Palaeolithic," *Cambridge Archaeological Journal* 25, no. 4 (2015).

30 *a New Zealand physician and adventurer:* David Lewis, *We the Navigators* (Honolulu: University Press, 1972).

32 *"They say you can tell the experienced navigators":* Thomas Gladwin, *East Is a Big Bird: Navigation and Logic on Puluwat Atoll* (Cambridge: Harvard University Press, 1970).

CHAPTER 2: CHASING DOPAMINE

36 *an oral history interview:* The Timoteo brothers' stories are told in *El Norte Es Como El Mar* (Guadalajara: Universidad de Guadalajara, 1995). Translations are available from the Mexican Migration Project website, mmp.opr.princeton.edu.

38 *about 75 percent heritable:* Stephen Faraone and Henrik Larsson, "Genetics of Attention Deficit Hyperactivity Disorder," *Molecular Psychiatry* 24 (2019).

39 *research with newborn babies:* for an overview see Jerome Kagan et al., "Reactivity in Infants: A Cross-National Comparison," *Developmental Psychology* 30, no. 3 (1994).

39 *geneticists at Yale published data:* F.M. Chang et al., "The world-wide distribution of allele frequencies at the human dopamine D4 receptor locus," *Human Genetics* 98, no. 1 (1996).

39 *a connection between DRD4 and hyperactivity:* G. J. LaHoste et al., "Dopamine D4 Receptor Gene Polymorphism Is Associated with Attention Deficit Hyperactivity Disorder," *Molecular Psychiatry* 1, no. 2 (1996).

39 *exploratory behavior in mice:* Stephanie Dulawa et al., "Dopamine D4 Receptor-Knock-Out Mice Exhibit Reduced Exploration of Novel Stimuli," *Journal of Neuroscience* 19, no. 21 (1999).

39 *linked DRD4 to novelty seeking:* Richard Ebstein et al., "Dopamine D4 Receptor (D4DR) Exon III Polymorphism Associated with the Human Personality Trait of Novelty Seeking," *Nature Genetics* 12, no. 1 (1996).

39 *Chen's idea was that the global distribution:* Chuansheng Chen et al., "Population Migration and the Variation of Dopamine D4 Receptor (DRD4) Allele Frequencies Around the Globe," *Evolution and Human Behavior* 20, no. 5 (1999).

40 *between forty thousand and fifty thousand years ago:* E. Wang et al., "The Genetic Architecture of Selection at the Human Dopamine Receptor D4 (DRD4) Gene Locus," *American Journal of Human Genetics* 74, no. 5 (2004).

41 *Katharine Montagu published her findings:* "Catechol Compounds in Rat Tissues and in Brains of Different Animals," *Nature* 180 no. 4579 (1957).

41 *Arvid Carlsson confirmed:* Arvid Carlsson et al., "On the Presence of 3-Hydroxytyramine in Brain," *Science* 127, no. 3296 (1958).

41 *it was a neurotransmitter:* Arvid Carlsson, "The Occurrence, Distribution and Physiological Role of Catecholamines in the Nervous System," *Pharmacological Reviews* 11, no. 2 (1959).

41 *"soups versus sparks"*: Elliot S. Valenstein, *The War of the Soups and the Sparks: The Discovery of Neurotransmitters and the Dispute over How Nerves Communicate* (New York: Columbia University Press, 2005).

41 *"But this was what I had insisted"*: Arvid Carlsson, "A Half-Century of Neurotransmitter Research: Impact on Neurology and Psychiatry," in *Nobel Lectures, Physiology or Medicine 1996-2000* (Singapore: World Scientific Publishing, 2003).

42 *the first experiments with L-dopa:* W. Birkmayer and O. Hornykiewicz, "Der L-Dioxyphenylalanin-Effekt bei der Parkinson-Akinese, *Wien Klinische Wochenschrift* 73 (1961); translated in Oleh Hornykiewicz, "Dopamine Miracle: From Brain Homogenate to Dopamine Replacement," *Movement Disorders* 17, no. 3 (2002).

42 *"sensory inputs are translated"*: Roy A. Wise, "The Dopamine Synapse and the Notion of 'Pleasure Centers' in the Brain," *Trends in Neurosciences* 3, no. 4 (1980).

42 *The first, published by Kent Berridge:* Kent Berridge et al., "Taste Reactivity Analysis of 6-Hydroxydopamine-Induced Aphagia: Implications for Arousal and Anhedonia Hypotheses of Dopamine Function," *Behavioral Neuroscience* 103, no. 1 (1989).

43 *linked dopamine to the concept of reward prediction error:* W. Schultz et al., "Responses of Monkey Dopamine Neurons to Reward and Conditioned Stimuli During Successive Steps of Learning a Delayed Response Task," *Journal of Neuroscience* 13, no. 3 (1993). The connection to reward prediction error came in W. Schultz et al., "A Neural Substrate of Prediction and Reward," *Science* 14, no. 5306 (1997).

43 *monkeys ramp up:* Vincent Costa et al., "Dopamine Modulates Novelty Seeking Behavior During Decision Making," *Behavioral Neuroscience* 128, no. 5 (2014).

44 *"a pronounced gain of function"*: Sergi Ferré et al., "Functional and Pharmacological Role of the Dopamine D4 Receptor and Its Polymorphic Variants," *Frontiers in Endocrinology* 13 (2022).

44 *bigger spike of reward signaling:* James Glazer et al., "DRD4 Polymorphisms Modulate Reward Positivity and P3a in a Gambling Task: Exploring a Genetic Basis for Cultural Learning," *Psychophysiology* 57, no. 10 (2020).

44 *network theory of migration:* Douglas Massey and Felipe García España, "The Social Process of International Migration," *Science* 237, no. 4816 (1987).

46 *"When he showed me the numbers"*: Jeffrey Napierala and Timothy Gage, "Social Networks and the Heritability of Migratory Behavior," *Biodemography and Social Biology* 62, no. 1 (2016).

46 *study of wild rhesus monkeys:* Sean P. Coyne et al., "Dopamine D4 Receptor Genotype Variation in Free-Ranging Rhesus Macaques and Its Association with Juvenile Behavior," *Behavioural Brain Research* 292 (2015).

46 *study of yellow-crowned bishops:* J. C. Mueller et al., "Behaviour-Related DRD4 Polymorphisms in Invasive Bird Populations," *Molecular Ecology* 23, no. 11 (2014).

46 *frog populations in mainland Sweden:* Tomas Brodin et al., "Personality Trait

Differences Between Mainland and Island Populations in the Common Frog (Rana temporaria)," *Behavioral Ecology and Sociobiology* 67, no. 1 (2013).

47 *House sparrows were introduced to Mombasa:* Andrea Liebl and Lynn Martin, "Living on the Edge: Range Edge Birds Consume Novel Foods Sooner Than Established Ones," *Behavioral Ecology* 25, no. 5 (2014).

47 *cane toads in Australia:* Benjamin Phillips et al., "Invasion and the Evolution of Speed in Toads," *Nature* 439, no. 7078 (2006); Gregory P. Brown et al., "The Straight and Narrow Path: the Evolution of Straight-Line Dispersal at a Cane Toad Invasion Front," *Proceedings of the Royal Society B: Biological Sciences* 281, no. 1795 (2014); Jodie Gruber et al., "Geographic Divergence in Dispersal-Related Behaviour in Cane Toads from Range-Front Versus Range-Core Populations in Australia," *Behavioral Ecology and Sociobiology* 71, no. 2 (2017).

48 *gene surfing:* Jayson Paulose and Oskar Hallatschek, "The Impact of Long-Range Dispersal on Gene Surfing," *PNAS* 117, no. 14 (2020).

48 *1.5 million Scandinavians:* Anne Sofie Beck Knudsen, "Those Who Stayed: Individualism, Self-Selection and Cultural Change During the Age of Mass Migration," *Social Science Research Network*, January 24, 2019.

48 *French-Canadian settlers:* Claudia Moreau et al., "Deep Human Genealogies Reveal a Selective Advantage to Be on an Expanding Wave Front," *Science* 334, no. 6059 (2011).

49 *the science journalist David Dobbs argues:* "Restless Genes," *National Geographic* 223, no. 1 (2013).

49 *led by Ron Pinhasi:* Ron Pinhasi et al., "Optimal Ancient DNA Yields from the Inner Ear Part of the Human Petrous Bone," *PLOS One* 10, no. 6 (2015).

49 *"Populations today almost never":* Chris Palmer, "The Skull's Petrous Bone and the Rise of Ancient Human DNA: Q & A with Genetic Archaeologist David Reich," *Biomedical Beat Blog—National Institute of General Medical Studies*, April 11, 2018.

49 *The British Isles, for example:* David Reich, *Who We Are and How We Got Here* (New York: Pantheon Books, 2018).

50 *extended Chen's findings:* Luke Matthews and Paul Butler, "Novelty-Seeking DRD4 Polymorphisms Are Associated with Human Migration Distance Out-of-Africa After Controlling for Neutral Population Gene Structure," *American Journal of Physical Anthropology* 145, no. 3 (2011).

50 *study led by Dan Eisenberg:* Dan Eisenberg et al., "Dopamine Receptor Genetic Polymorphisms and Body Composition in Undernourished Pastoralists: An Exploration of Nutrition Indices Among Nomadic and Recently Settled Ariaal Men of Northern Kenya," *BMC Ecology and Evolution* 173 (2008).

CHAPTER 3: THE FREE ENERGY PRINCIPLE

53 *Two paths diverged:* I told this story in the June 2012 issue of *Men's Journal*, "No Path, No Rules: The Wildest New Race."

54 *"If you want to be a champion":* quoted in Tim Noakes, *Lore of Running* (Champaign: Leisure Press, 1991).

54 *perception of time depends in part:* David Behm and Tori Carter, "Effect of Exercise-Related Factors on the Perception of Time," *Frontiers in Physiology* 11 (2020).

55 *"Compared to a more experienced orienteer":* quoted in Alex Hutchinson, "No Path, No Rules."

55 *the general term I'll use here is "predictive processing":* Andy Clark, "Whatever next? Predictive brains, situated agents, and the future of cognitive science," *Behavioral and Brain Science* 36, no. 3 (2013). For a broader overview of predictive processing, see Andy Clark, *The Experience Machine: How Our Minds Predict and Shape Reality* (New York: Pantheon Books, 2023).

56 *"a strange architecture":* Patrick Winston, "The Next 50 Years: A Personal View," *Biologically Inspired Cognitive Architectures* 1 (2012).

56 *Rajesh Rao and Dana Ballard presented:* "Predictive Coding in the Visual Cortex: A Functional Interpretation of Some Extra-Classical Receptive-Field Effects," *Nature Neuroscience* 2, no. 1 (1999).

57 *an unremarkable dress:* Pascal Wallisch, "Illumination Assumptions Account for Individual Differences in the Perceptual Interpretation of a Profoundly Ambiguous Stimulus in the Color Domain: 'The dress,'" *Journal of Vision* 17, no. 5 (2017).

57 *Karl Friston published an attempt:* "A Theory of Cortical Responses," *Philosophical Transactions of the Royal Society B* 360, no. 1456 (2005). See also Friston et al., "A Free Energy Principle for the Brain," *Journal of Physiology-Paris* 100 (2006).

57 *"understanding everything":* quoted in an excellent profile of Friston and his work by Shaun Raviv, "The Man Who Explained Everything," *Wired* 26, no. 12 (2018).

57 *"the negative log-probability":* Karl Friston, "The Free-Energy Principle: A Rough Guide to the Brain?" *Trends in Cognitive Science* 13, no. 7 (2009).

58 *a mathematical statement:* If you'd like to dig into the math and concepts underlying the free energy principle, try Thomas Parr, Giovanno Pezzulo, and Karl Friston's book *Active Inference: The Free Energy Principle in Mind, Brain, and Behavior* (Cambridge: MIT Press, 2022), which is available as a free ebook from MIT Press. It's a relatively accessible introduction (with the emphasis on "relatively").

58 *"There was a lot of mathematical knowledge":* Peter Freed, "Research Digest," *Neuropsychoanalysis* 12, no. 1 (2010).

58 *how the brain is structured:* Karl Friston and Klaas Stephan, "Free-Energy and the Brain," *Synthese* 159, no. 3 (2007).

58 *psychiatric disorders such as depression:* Paul Badcock et al., "The Depressed Brain: An Evolutionary Systems Theory," *Trends in Cognitive Science* 21, no. 3 (2017).

58 *enjoy horror movies:* Mark Miller et al., "Surfing Uncertainty with Screams: Predictive Processing, Error Dynamics and Horror Films," *Philosophical Transactions of the Royal Society B* 379, no. 1895 (2024).

58 *growth of cancer cells:* Noor Sajid et al., "Cancer Niches and Their Kikuchi Free Energy," *Entropy* 23, no. 5 (2021).

59 *the Dark Room problem:* Karl Friston et al., "Free-Energy Minimization and the Dark-Room Problem," *Frontiers in Psychology* 3 (2012).

60 *a team led by Rosalyn Moran:* Maell Cullen et al., "Active Inference in OpenAI Gym: A Paradigm for Computational Investigations into Psychiatric Illness," *Biological Psychiatry: Cognitive Neuroscience and Neuroimaging* 3, no. 9 (2018). See also the description in Raviv, "The Man Who Explained Everything."

61 *asked volunteers to sit quietly in a room:* Timothy Wilson et al., "Just Think: The Challenges of the Disengaged Mind," *Science* 345, no. 6192 (2014).

62 *follow-up study with 254 participants in Germany:* Andreas Eder et al., "Motivations Underlying Self-Infliction of Pain During Thinking for Pleasure," *Scientific Reports* 12 (2022).

62 *"the mind is designed to engage":* Fariss Samarrai, "Doing Something Is Better Than Doing Nothing for Most People, Study Shows," *UVAToday*, July 3, 2014.

62 *In Hsee and Ruan's study:* Christopher Hsee and Bowen Ruan, "The Pandora Effect: The Power and Peril of Curiosity," *Psychological Science* 27, no. 5 (2016); see also Ruan et al., "The Teasing Effect: An Underappreciated Benefit of Creating and Resolving an Uncertainty," *Journal of Marketing Research* 55, no. 4 (2018).

62 *a restaurant in Sydney:* The restaurant was called Wagaya. Its successor, Lantern by Wagaya, still offers sushi roulette as of 2024.

63 *"Sometimes knowledge is sought":* Zachary Wojtowicz, Nick Chater, and George Loewenstein, "The Motivational Processes of Sense-Making," in *The Drive for Knowledge: The Science of Human Information Seeking*, eds. Irene Cogliati Dezza, Eric Schulz, and Charley M. Wu (Cambridge: Cambridge University Press, 2022).

63 *"the feeling of grip":* Julian Kiverstein et al., "The Feeling of Grip: Novelty, Error Dynamics, and the Predictive Brain," *Synthese* 196, no. 7 (2019).

64 *"slope-chasing":* quoted in Kiverstein, "The Feeling of Grip."

64 *Play, according to a theory:* Marc Malmdorf Andersen et al., "Play in Predictive Minds: A Cognitive Theory of Play," *Psychological Review* 130, no. 2 (2023).

65 *sometimes called the Wundt Curve:* see, e.g., Sebastian Deterding et al., "Mastering Uncertainty: A Predictive Processing Account of Enjoying Uncertain Success in Video Game Play," *Frontiers in Psychology* 13 (2022). Earlier discussion in Daniel Berlyne, *Conflict, Arousal, and Curiosity* (New York: McGraw-Hill, 1960), citing Wilhelm Wundt's masterwork *Grundzüge der physiologischen Psychologie* (Leipzig: Engelmann, 1874).

66 *games like peekaboo:* W. Gerrod Parrott and Henry Gleitman, "Infants' Expectations in Play: The Joy of Peek-a-boo," *Cognition and Emotion* 3, no. 4 (1989).

66 *"Goldilocks effect":* Celeste Kidd et al., "The Goldilocks Effect: Human Infants Allocate Attention to Visual Sequences That Are Neither Too Simple nor Too Complex," *PLOS ONE* 7, no. 5 (2012).

67 *My very first magazine assignment:* the RoboCup story ran on the web, but (as far as I can tell) is no longer available online.

67 *RoboCup was established in 1997:* For more on RoboCup, see www.robocup.org.

67 *"That was actually a whole adventure":* Frédéric Kaplan, oral history conducted in 2011 by Peter Asaro, Indiana University and the IEEE History Center. The IEEE's Engineering and Technology History Wiki has a collection of more than 800 oral histories in electrical and computer technology which can be accessed at ethw.org .

68 *"the only known process":* Anna Giron et al., "Developmental Changes in Exploration Resemble Stochastic Optimization," *Nature Human Behaviour* 7, no. 11 (2023).

68 *Videos of Oudeyer and Kaplan's experiments:* "Intrinsically Motivated Learning and Curiosity in The Playground Experiment (AIBO robot)," Feb. 3, 2011, https://youtu.be/uAoNzHjzzys.

69 *Oudeyer and Kaplan published the details:* Pierre-Yves Oudeyer et al., "The Playground Experiment: Task-Independent Development of a Curious Robot," *Proceedings of the AAAI Spring Symposium on Developmental Robots* (San Francisco: AAAI, 2005).

69 *dopamine and other neuromodulators may play a broader role:* for a recent overview, see Kelly Diederen and Paul Fletcher, "Dopamine, Prediction Error and Beyond," *Neuroscientist* 27, no. 1 (2021).

70 *the cognitive benefits of orienteering:* Emma Waddington et al., "Orienteering Combines Vigorous-Intensity Exercise with Navigation to Improve Human Cognition and Increase Brain-Derived Neurotrophic Factor," *PLOS ONE* 19, no. 5 (2024).

70 *as the Israeli neuroscientist Moshe Bar puts it:* in *Mindwandering: How Your Constant Mental Drift Can Improve Your Mood and Boost Your Creativity* (New York: Hachette, 2022).

71 *Bar, with his colleague Noa Herz:* Noa Herz et al., "Overarching States of Mind," *Trends in Cognitive Sciences* 24, no. 3 (2020).

71 *his book on the pleasures of suffering:* Paul Bloom, *The Sweet Spot: Suffering, Pleasure and the Key to a Good Life* (New York: Ecco, 2021).

CHAPTER 4: EXPLORE VS. EXPLOIT

75 *On May 3, 2019, Nils van der Poel:* dates and details of van der Poel's training and racing are primarily from the manifesto he published after the Beijing Olympics, "How to Skate a 10k . . . and Also Half a 10K," available at www.howtoskate.se. Other details of his background are from his interview with Stephen Seiler: "Reward and Forgiveness: Discussing Training with Champion Speed Skater Nils van der Poel," March 21, 2002, https://youtu.be/eZ76s-5_F70.

76 *This terminology was first proposed:* J. C. Gittins and D. M. Jones, "A Dynamic Allocation Index for New-Product Chemical Research," University of Cambridge Engineering Department (1974), cited in John Gittins et al., *Multi-Armed Bandit Allocation Indices* (Chichester: John Wiley & Sons, 2011). This is

a fairly obscure reference; the explore-exploit terminology seems to have come into more common usage after James March's paper "Exploration and Exploration in Organizational Learning," *Organizational Science* 2, no. 1 (1991).

77 *a thousand skydives:* Kevin Draper, "These Might Be the Best Quotes About Motivation You'll Hear in These Olympics," *New York Times*, February 5, 2022.

78 *"Man, I've never seen anyone like him before":* translated from "Jag Har Aldrig Sett Någon Som van der Poel [I've Never Seen Anyone like van der Poel Before]," *TT-Sportbladet*, November 11, 2021.

78 *"Some said my plan was foolish":* quoted in "Basic Military Training Re-ignited Skate Star Nils van der Poel's Career," *Swedish Armed Forces Headquarters*, February 16, 2021, www.forsvarsmakten.se.

79 *Robert Bush and Frederick Mosteller:* "A Stochastic Model with Applications to Learning," *Annals of Mathematical Statistics* 24, no. 4 (1953); cited in Tor Lattimore and Csaba Szepesvàri, *Bandit Algorithms* (Cambridge: Cambridge University Press, 2020).

80 *Peter Whittle quipped:* J. C. Gittins, "Bandit Processed and Dynamic Allocation Indices," *Journal of the Statistical Society—Series B (Methodological)* 41, no. 2 (1979).

80 *a solution had been found:* for an account of the development of the Gittins index, see Brian Christian and Tom Griffiths, *Algorithms to Live By: The Computer Science of Human Decisions* (New York: Henry Holt, 2016).

80 *Gittins's 1989 textbook:* or you could consult the updated second edition, Gittins et al., *Multi-Armed Bandit Allocation Indices* (Chichester: John Wiley & Sons, 2011). The Gittins indices I cite for Alice's and Bob's restaurants assume a "geometric discounting" factor of 0.9: that is, I consider tomorrow's dinner only 90 percent as important as tonight's.

81 *two broad categories of exploring strategy:* Samuel Gershman, "Deconstructing the Human Algorithms for Exploration," *Cognition* 173 (2018); Robert Wilson et al., "Balancing Exploration and Exploitation with Information and Randomization," *Current Opinion in Behavioral Sciences* 38 (2021).

82 *When some choices are much more uncertain than others:* Samuel Gershman and Naoshige Uchida, "Believing in Dopamine," *Nature Reviews Neuroscience* 20, no. 11 (2019).

83 *random variability in the signals:* Samuel Feng et al., "The Dynamics of Explore–Exploit Decisions Reveal a Signal-to-Noise Mechanism for Random Exploration," *Scientific Reports* 11 (2021).

83 *"There's a piece of the brain":* quoted in Ewen Calloway, "Random Babbling Leads Chicks to the Perfect Tune," *New Scientist*, May 2, 2008; see also Dmitriy Arnov et al., "A Specialized Forebrain Circuit for Vocal Babbling in the Juvenile Songbird," *Science* 320, no. 5876 (2008).

83 *continue to harness randomness:* Jonnathan Singh Alvarado et al., "Neural Dynamics Underlying Birdsong Practice and Performance," *Nature* 599 (2021).

83 *"After a two-year hiatus":* translated from "Efter succén—van der Poel tror på OS-medalj [After the success—van der Poel believes in an Olympic medal]," *TT*

Nyhetsbyrån, January 17, 2021. In the same article, he cites Therese Johaug as an inspiration.

86 *food delivery company Deliveroo*: Eric Schulz et al., "Structured, Uncertainty-Driven Exploration in Real-World Consumer Choice," *PNAS* 116, no. 28 (2019).

88 *"optimism in the face of uncertainty"*: Christian and Griffiths, in *Algorithms to Live By*, cite Leslie Kaelbling et al., "Reinforcement Learning: A Survey," *Journal of Artificial Intelligence Research* 4 (1996), as a source for this phrase.

88 *Tze Leung Lai and Herbert Robbins showed*: "Asymptotically Efficient Adaptive Allocation Rules," *Advances in Applied Mathematics* 6, no. 1 (1985).

89 *Longline fishing boats*: Shay O'Farrell et al., "Disturbance Modifies Payoffs in the Explore-Exploit Trade-off," *Nature Communications* 10 (2019).

89 *London Tube workers began a forty-eight-hour strike*: Shaun Larcom et al., "The Benefits of Forced Experimentation: Striking Evidence from the London Underground Network," *The Quarterly Journal of Economics* 132, no. 4 (2017).

90 *"The walk from Liverpool Street"*: quoted in Larcom, "Forced Experimentation."

90 *45 percent of the things we do*: Wendy Wood and David Neal, "The Habitual Consumer," *Journal of Consumer Psychology* 19 (2009).

90 *Herbert Simon coined the term*: "A Behavioral Model of Rational Choice," *Quarterly Journal of Economics* 69, no. 1 (1955).

91 *the Porter Hypothesis*: Michael Porter, "America's Green Strategy," *Scientific American* 264, no. 4 (1991).

91 *a scenario that Merrill Flood presented*: Thomas Ferguson, "Who Solved the Secretary Problem?" *Statistical Science* 4, no. 3 (1989). His daughter's role is recounted in a letter from Flood to Ferguson, dated May 12, 1988, held by the Bentley Historical Library at the University of Michigan in the Merrill M. Flood Papers; it is also told in Christian and Griffiths, *Algorithms to Live By*.

92 *a precise and provably optimal solution*: Ferguson's "Who Solved the Secretary Problem?" has a detailed discussion of who should get credit for the 37 percent rule. The problem first appeared in print in Martin Gardner's *Scientific American* column in 1960, with a subsequent solution attributed to Leo Moser and J. R. Pounder. But Merrill had written a letter in 1958 that contained the solution, according to Christian and Griffiths in *Algorithms to Live By*, and he thought others had solved it too.

92 *the 37 percent rule*: see, e.g., Andrew Trees, *Decoding Love* (New York: Avery, 2009).

92 *a study of job-hiring decisions*: Darryl Seale and Amnon Rapoport, "Sequential Decision Making with Relative Ranks: An Experimental Investigation of the 'Secretary Problem,'" *Organizational Behavior and Human Decision Processes* 69, no. 3 (1997).

93 *Kepler famously spent two years*: recounted in Ferguson, "Who Solved the Secretary Problem?"

94 *"I have frequently used the algorithm"*: from Flood's letter to Ferguson, May 12, 1988.

NOTES 261

94 *"He Broke Every Record. Then He Told His Rivals How to Beat Him"*: Joshua Robinson, *Wall Street Journal*, February 13, 2022.

94 *Trail Runner Magazine lauded*: David Roche, "The Wildly Cool Training Approach of Speed Skating Gold Medalist Nils van der Poel," updated February 14, 2022.

94 *"There probably isn't an endurance athlete in Sweden"*: translated from Ida Forsgren, "Nils van der Poel–Effekt Bakom Andreas Almgrens Rekordform [Nils van der Poel Effect Behind Andreas Almgren's record form]," *SVT Sport*, September 19, 2022.

94 *"working beyond all expectations"*: "The Van der Poel's Training Method Is Working Beyond All Expectations," *ProXCSkiing.com*, October 21, 2022.

94 *"I've been trying to quit"*: Draper, *New York Times*.

94 *upped his odds of retirement to 80 percent*: translated from Luuk Blijboom, "Van der Poel Bereikt Bestemming, met Dank aan Omstreden Gids [Van der Poel Reaches Destination, Thanks to a Controversial Guide]," *NOS*, February 11, 2022.

95 *"As soon as you go in one direction"*: translated from interview with *Radiosporten*, February 8, 2022, twitter.com/Radiosporten/status/1491090244652969984.

95 *"If he wants, Nils could get in shape"*: translated from Mats Bråstedt, "Van der Poels Tränare: 'Då Kan Han Göra Comeback' [Van der Poel's coach: 'Then he can make a comeback']," *Expressen*, September 28, 2022.

CHAPTER 5: UNKNOWN UNKNOWNS

96 *After yet another sixteen-hour day of canoeing*: the account of Alexander Mackenzie's voyages is drawn primarily from *The Journals and Letters of Sir Alexander Mackenzie*, ed. W. Kaye Lamb (Cambridge: Cambridge University Press, 1970), as well as Derek Hayes, *First Crossing: Alexander Mackenzie, His Expedition Across North America, and the Opening of a Continent* (Vancouver: Douglas & McIntyre, 2001).

97 *three thousand miles by canoe to get back to Montreal*: Eric Morse, *Fur Trade Routes of Canada: Then and Now* (Ottawa: Queen's Printer, 1969).

97 *six days' travel away*: Peter Pond, Mackenzie's fur-trading mentor and predecessor at Lake Athabasca, told the president of Yale in 1790 that he had been within six days of the Pacific coast, as noted in Lamb's introduction to Mackenzie's journals.

99 *a twenty-two-year-old named Tyler Bradt*: trailer for documentary *Dream Result*, Rush Sturges and Tyler Bradt, August 21, 2010, https://youtu.be/uNXh9gXDd2Y.

99 *Bradt broke his back*: "The Perils of Big Plunges: Jesse Coombs Collapses Lung on Abiqua, Tyler Bradt Healing from Broken Back," *Paddling Life*, May 25, 2011.

99 *In a Pentagon news conference*: For a fascinating deconstruction of Rumsfeld's "known unknowns," see Errol Morris's four-part essay "The Certainty of Donald Rumsfeld," *New York Times*, March 25, 26, 27, and 28, 2014.

99 *Both John Keats and Robert Browning*: cited in Morris, "The Certainty of Donald Rumsfeld."

99 *John Maynard Keynes and Frank Knight:* for an overview, see Mark Packard et al., "Keynes and Knight on Uncertainty: Peas in a Pod or Chalk and Cheese?" *Cambridge Journal of Economics* 45, no. 5 (2021).

100 *"the most dangerous man in America":* this account of Ellsberg's life and work is drawn primarily from Steve Sheinkin, *Most Dangerous: Daniel Ellsberg and the Secret History of the Vietnam War* (New York: Roaring Brook, 2015); Daniel Ellsberg, *Secrets: A Memoir of Vietnam and the Pentagon Papers* (New York: Penguin, 2003); Daniel Ellsberg, *The Doomsday Machine: Confessions of a Nuclear War Planner* (New York: Bloomsbury USA, 2017).

100 *Ellsberg's dissertation: Risk, Ambiguity and Decision* (New York: Garland, 2001).

101 *"subjective uncertainty when experience":* Ellsberg, *The Doomsday Machine.*

101 *In a 1961 paper:* Daniel Ellsberg, "Risk, Ambiguity, and the Savage Axioms," *Quarterly Journal of Economics* 75, no. 4 (1961). This is the classic presentation of the Ellsberg paradox.

101 *researchers at Yale offered volunteers:* Ruonan Ji et al., "Learning About the Ellsberg Paradox Reduces, but Does Not Abolish, Ambiguity Aversion," *PLOS ONE* 15, no. 3 (2020).

102 *separate neural circuitry:* Shuyi Wu et al., "Better the Devil You Know Than the Devil You Don't: Neural Processing of Risk and Ambiguity," *NeuroImage* 236 (2021).

102 *"It would be the transcendent":* Ellsberg, *The Doomsday Machine.* Ellsberg's recollections of Cheyenne Mountain and the Tonkin Gulf are also drawn from this book.

104 *River Disappointment:* Mackenzie used this name, presumably tongue-in-cheek, in a subsequent letter to his cousin. It was never officially proposed or marked on a map.

105 *rather than 135 degrees west:* Hayes, *First Crossing.*

105 *more than three thousand miles:* W. Kaye Lamb, "Mackenzie, Sir Alexander," in *Dictionary of Canadian Biography*, vol. 5 (Toronto: University of Toronto Press, 1983).

106 *George Loewenstein points out:* "Because It Is There: The Challenge of Mountaineering . . . for Utility Theory," *Kyklos* 52, no. 3 (1999).

106 *"the blessings of the government":* Alfred Lansing, *Endurance: Shackleton's Incredible Voyage* (New York: Carroll & Graf, 1959), quoted in Loewenstein, "Because It Is There."

107 *Daniel Berlyne's proposed motivations:* D. E. Berlyne, *Conflict, Arousal, and Curiosity* (New York: McGraw-Hill, 1960).

108 *the Horizon Task:* Robert Wilson et al., "Humans Use Directed and Random Exploration to Solve the Explore–Exploit Dilemma," *Journal of Experimental Psychology: General* 143, no. 6 (2014).

109 *Lapidow and Elizabeth Bonawitz published:* "What's in the Box? Preschoolers Consider Ambiguity, Expected Value, and Information for Future Decisions in Explore-Exploit Tasks," *Open Mind* 7 (2023).

111 *"As he used to indulge himself"*: quoted in Hayes, *First Crossing*.
112 *inscribed these words:* This is the wording recorded in Mackenzie's journal. What is now carved into the rock is "Alex Mackenzie from Canada by land 22d July 1793."
113 *almost certainly carrying a copy of Mackenzie's book:* Arlen J. Large, "Mackenzie's Wonderful Trail to Nowhere," *We Proceeded On* 19, no. 4 (1993), available at lewis-clark.org.

CHAPTER 6: THE TIME HORIZON

114 *In a limestone cave:* I wrote about Ella Shpayer's research ("Not Just Child's Play") in the Fall 2022 issue of *Aperio* magazine; for details see Ella Assaf, "Dawn of a New Day: The Role of Children in the Assimilation of New Technologies Throughout the Lower Paleolithic," *L'anthropologie* 125, no. 1 (2021).
115 *Jean Twenge and Heejung Park published:* "The Decline in Adult Activities Among U.S. Adolescents, 1976–2016," *Child Development* 90, no. 2 (2019).
115 *"Longer Adolescence Is Held Key": New York Times*, January 3, 1971.
115 *Charles Dickens himself:* Michael Allen, "New Light on Dickens and the Blacking Factory," *The Dickensian* 106, no. 480 (2010).
116 *Chickens are born:* Alexander McNamara, "Chicken, Run! Newborn Chicks Born Able to 'Recognise and React' to Dangers," *BBC Science Focus*, October 29, 2019.
116 *Horses can get up:* "Are Our Big Brains the Reason Newborns Can't Walk?" *Scientific American*, September 1, 2009.
116 *chimpanzees can gather:* Hillard Kaplan et al., "A Theory of Human Life History Evolution: Diet, Intelligence, and Longevity," *Evolutionary Anthropology* 9, no. 4 (2009).
116 *Alison Gopnik advanced a more utilitarian hypothesis:* for an overview, see "Childhood as a Solution to Explore–Exploit Tensions," *Philosophical Transactions of the Royal Society B: Biological Sciences* 375, no. 1803 (2020).
117 *"blicket detector":* Christopher Lucas et al., "When Children Are Better (or at Least More Open-Minded) Learners Than Adults: Developmental Differences in Learning the Forms of Causal Relationships," *Cognition* 131 (2014).
117 *wooden blocks were "zaffs":* Emily Liquin and Alison Gopnik, "Children Are More Exploratory and Learn More Than Adults in an Approach-Avoid Task," *Cognition* 218 (2022).
118 *Anna Giron led a 2023 study:* Anna Giron et al., "Developmental Changes in Exploration Resemble Stochastic Optimization," *Nature Human Behaviour* 7 (2023).
119 *Richard Bellman worked out the details:* "A Problem in the Sequential Design of Experiments," *Sankhya* 16, no. 3/4 (1956).
119 *Horizon Task performance of people in their sixties and seventies:* Jack-Morgan Mizell, "Horizon Task Performance of People in Their 60s and 70s," *Psychology and Aging* 39, no. 1 (2024).
120 *A 2022 analysis by researchers in Switzerland:* Johannes Burtscher et al., "The

Impact of Training on the Loss of Cardiorespiratory Fitness in Aging Masters Endurance Athletes," *International Journal of Environmental Research and Public Health* 19, no. 17 (2022).

120 *about half as many synaptic connections:* Zdravko Petanjek et al., "Extraordinary Neoteny of Synaptic Spines in the Human Prefrontal Cortex," *PNAS* 108, no. 32 (2011).

121 *In 2022, all ten of the highest-grossing Hollywood hits:* as compiled by Box Office Mojo, www.boxofficemojo.com.

121 *"The movie business as before":* quoted in David Gura, "Barry Diller Headed 2 Hollywood Studios. He Now Says the Movie Business Is Dead," *NPR*, July 8, 2021.

122 *life expectancy for babies:* period life expectancies from "The 2024 Annual Report of the Board of Trustees of the Federal Old-Age and Survivors Insurance and Federal Disability Insurance Trust Funds," available at www.ssa.gov.

122 *three times more likely to report having changed jobs:* Brandon Rigoni and Amy Adkins, "Millennials Job-Hoppers: What They Seek," *Gallup*, May 19, 2016, news.gallup.com.

123 *"Chinese music":* quoted in Krin Gabbard, *Hotter Than That: The Trumpet, Jazz, and American Culture* (New York: Farrar, Straus and Giroux: 2015).

123 *"weird chords which don't mean nothing":* quoted in Terry Teachout, "Jazz as Modern Art," *Commentary*, January 2023.

123 *Andersen's theory of play:* Marc Malmdorf Andersen et al., "Play in Predictive Minds: A Cognitive Theory of Play," *Psychological Review* 130, no. 2 (2023).

124 *our musical preferences follow a Wundt curve:* Benjamin Gold et al., "Predictability and Uncertainty in the Pleasure of Music: A Reward for Learning?" *The Journal of Neuroscience* 39, no. 47 (2019).

124 *Igor Stravinsky's The Rite of Spring:* the controversial debut is described in Thomas Forrest Kelly, *First Nights: Five Musical Premiers* (New Haven: Yale University Press, 2000); and Ivan Hewitt, "The Rite of Spring 1913: Why Did It Provoke a Riot?" *The Telegraph*, May 16, 2013.

124 *Alex Ross recounts: The Rest Is Noise: Listening to the Twentieth Century* (New York: Farrar, Straus and Giroux, 2007).

125 *"Music is like wine, Bridgetower":* Robertson Davies, *A Mixture of Frailties* (Toronto: Macmillan, 1958).

125 *journalist Tim Falconer:* "No City for Middle-Aged Men," *Toronto Standard*, August 17, 2011.

126 *age of peak musical exploration:* Callum Davies et al., "The Power of Nostalgia: Age and Preference for Popular Music," *Marketing Letters* 33 (2022).

126 *"Most people have all the songs":* Jeremy Larson, "Why Do We Even Listen to New Music?" *Pitchfork*, April 6, 2020.

128 *exploring patterns of patients with schizophrenia:* James Waltz et al., "Differential Effects of Psychotic Illness on Directed and Random Exploration," *Computational Psychiatry* 4 (2020).

NOTES

CHAPTER 7: MAPPING THE WORLD

130 *Robert O'Hara Burke was the first:* my account of the Burke and Wills expedition is drawn primarily from David Phoenix, *"More Like a Picnic Party": Burke and Wills: An Analysis of the Victorian Exploring Expedition of 1860-1861*, PhD thesis, James Cook University (2017); and Peter Fitzsimons, *Burke & Wills: The Triumph and Tragedy of Australia's Most Famous Explorers* (Sydney: Hachette, 2017). I have also drawn on the original archival material (in particular the Commission of Enquiry) made available at David Phoenix's Burke & Wills Web, www.burkeandwills.net.au. In cases where sources conflict, my unwavering rule was "Always go with Phoenix."

131 *blazes on two coolabah trees:* exactly which tree had what carved on it is up for debate: Maddelin McCosker, "Researcher Casts Doubt on Legitimacy of Burke and Wills Dig Tree," *ABC News*, October 12, 2017.

132 *the redesign of Ciudad Guayana:* Donald Appleyard, "Styles and Methods of Structuring a City," *Environment and Behavior* 2, no. 1 (1970).

132 *Kids who walk to school:* Antonella Rissotto and Francesco Tonucci, "Freedom of Movement and Environmental Knowledge in Elementary School Children," *Journal of Environmental Psychology* 22, no. 1-2 (2002); cited in Michael Bond, *From Here to There: The Art and Science of Finding and Losing Our Way* (Cambridge: Harvard University Press: 2020), which provides an exceptional overview of the science of wayfinding, as does M. R. O'Connor's *Wayfinding: The Science and Mystery of How Humans Navigate the World* (New York: St. Martin's Press, 2019).

132 *Willard Small, of Clark University in Massachusetts, published the results:* "Experimental Study of the Mental Processes of the Rat. II," *American Journal of Psychology* 12, no. 2 (1901).

133 *according to behavioral neuroscientist Paul Dudchenko:* in *Why People Get Lost: The Psychology and Neuroscience of Spatial Cognition* (Oxford: Oxford University Press, 2010).

133 *a 1948 rats-in-mazes paper:* Edward Tolman, "Cognitive Maps in Rats and Men," *Psychological Review* 55, no. 4 (1948).

133 *John Watson and B. F. Skinner's behaviorist theories:* for an accessible (and entertaining) overview, see Paul Bloom, *Psych: The Story of the Human Mind* (New York: Ecco, 2023).

134 *a behaviorist named Karl Lashley: Brain Mechanisms and Intelligence: A Quantitative Study of Injuries to the Brain* (Chicago: University of Chicago Press, 1929).

134 *The eventual breakthrough:* the discovery of place cells is recounted in John O'Keefe's biography in *The Nobel Prizes 2014* (Sagamore Beach: Science History Publications, 2015).

135 *"grid cells":* Torkel Hafting et al., "Microstructure of a Spatial Map in the Entorhinal Cortex," *Nature* 436 (2005).

135 *"head direction cells":* J. S. Taube et al., "Head-Direction Cells Recorded from the Postsubiculum in Freely Moving Rats. I. Description and Quantitative Analysis," *Journal of Neuroscience* 10, no. 2 (1990).

135 *"boundary cells"*: Colin Lever et al., "Boundary Vector Cells in the Subiculum of the Hippocampal Formation," *Journal of Neuroscience* 29, no. 31 (2009).
135 *"doorway effect"*: Kyle Pettijohn and Gabriel Radvansky, "Walking Through Doorways Causes Forgetting: Event Structure or Updating Disruption?" *Quarterly Journal of Experimental Psychology* 69, no. 11 (2016).
135 *Willem Janszoon:* "Janszoon Maps Northern Australian Coast," *National Museum of Australia*, www.nma.gov.au.
136 *Jane Jacobs observed:* in *The Death and Life of Great American Cities* (New York: Random House, 1961).
136 *Anxious mice spend more time:* P. Simon et al., "Thigmotaxis as an Index of Anxiety in Mice. Influence of Dopaminergic Transmissions," *Behavioural Brain Research* 61, no. 1 (1994).
136 *"high anxiety sensitivity"*: Nora Walz et al., "A Human Open Field Test Reveals Thigmotaxis Related to Agoraphobic Fear," *Biological Psychiatry* 80, no. 5 (2016).
136 *moving one wall to stretch a square room:* John O'Keefe and Neil Burgess, "Geometric Determinants of the Place Fields of Hippocampal Neurons," *Nature* 381 (1996).
137 *boundary cells were identified in rat brains:* Colin Lever et al., "Boundary Vector Cells in the Subiculum of the Hippocampal Formation," *Journal of Neuroscience* 29, no. 31 (2009).
137 *mythical Great Inland Sea:* see, for example, Thomas Maslen's 1827 map, as shown in Frank Jacobs, "The Great Australian Inland Sea," *Big Think*, July 23, 2010, bigthink.com.
138 *nine hundred million acres:* from *Argus*, November 20, 1857, cited in Phoenix, "More Like a Picnic Party."
138 *Morris's original 1981 study:* Richard G. Morris, "Spatial Localization Does Not Require the Presence of Local Cues," *Learning and Motivation* 12, no. 2 (1981).
138 *eight distinct exploring strategies:* Tiago Gehring et al., "Detailed Classification of Swimming Paths in the Morris Water Maze: Multiple Strategies Within One Trial," *Scientific Reports* 5 (2015).
139 *German researchers pitted eighteen mice:* Robby Schoenfeld et al., "Variants of the Morris Water Maze Task to Comparatively Assess Human and Rodent Place Navigation," *Neurobiology of Learning and Memory* 139 (2017).
139 *more than two thousand miles away:* Phoenix puts the total one-way distance traveled by the expedition at 3,700 kilometers (2,300 miles). A crow would cover considerably less distance.
139 *Sarah Murgatroyd asked in her influential 2002 account: The Dig Tree: A True Story of Bravery, Insanity, and the Race to Discover Australia's Wild Frontier* (New York: Broadway, 2002).
144 *a striking pattern in the location:* Ariane Burke, "Spatial Abilities, Cognition and the Pattern of Neanderthal and Modern Human Dispersals," *Quaternary International* 247 (2012).

145 *London taxi drivers:* Eleanor Maguire et al., "Navigation-Related Structural Change in the Hippocampi of Taxi Drivers," *PNAS* 97, no. 8 (2000).
145 *volunteers found their way around in a virtual reality world:* G. Iaria et al., "Cognitive Strategies Dependent on the Hippocampus and Caudate Nucleus in Human Navigation: Variability and Change with Practice," *Journal of Neuroscience* 23, no. 13 (2003).
145 *84 percent of children:* Véronique Bohbot et al., "Virtual Navigation Strategies from Childhood to Senescence: Evidence for Changes Across the Life Span," *Frontiers in Aging Neuroscience* 4 (2012).

CHAPTER 8: THE LANDSCAPE OF IDEAS

149 *Oxford flu:* Rolf Landauer, quoted in Ivars Peterson, *Science News* 147, no. 2 (1995).
149 *"the creation and interpretation of new knowledge":* cited as one of the required learning outcomes in Cambridge's course directory for "PhD in Physics," https://www.postgraduate.study.cam.ac.uk/courses/directory/pcphpdphy.
149 *"Voyaging through strange seas":* William Wordsworth, *The Prelude, or Growth of a Poet's Mind* (London: Edward Moxon, 1850).
150 *a tradition started by J. J. Thomson:* P. Phillips, "Sir Joseph John Thomson, the Great English Physicist," *Scientific American* 105, no. 26 (1911).
150 *a famous 1981 lecture:* Richard Feynman, "Simulating Physics with Computers," *International Journal of Theoretical Physics* 21, no. 6/7 (1982).
150 *David Deutsch drew up a formal definition:* "Quantum Theory, the Church–Turing Principle and the Universal Quantum Computer," *Proceedings of the Royal Society A: Mathematical, Physical and Engineering Sciences* 400, no. 1818 (1985).
150 *Peter Shor published an algorithm:* "Algorithms for Quantum Computation: Discrete Logarithms and Factoring," *Proceeding of the 35nd Annual Symposium on Foundations of Computer Science,* (Santa Fe: IEEE, 1994).
151 *"The stuff that I did in the late nineteen-seventies":* quoted in Rivka Galchen, "Dream Machine," *The New Yorker,* May 2, 2011.
152 *keep track of our social networks:* Rita Morais Tavares et al., "A Map for Social Navigation in the Human Brain," *Neuron* 87, no. 1 (2015).
152 *Benjamin Pitt and Daniel Casasanto point out:* "Spatial Metaphors and the Design of Everyday Things," *Frontiers in Psychology* 13 (2022).
152 *"QWERTY effect":* Kyle Jasmin and Daniel Casasanto, "The QWERTY Effect: How Typing Shapes the Meanings of Words," *Psychonomic Bulletin & Review* 19 (2012).
153 *Howard Eichenbaum trained rats:* M. Bunsey and H. Eichenbaum, "Conservation of Hippocampal Memory Function in Rats and Humans," *Nature* 379, no. 6562 (1996).
153 *A 2020 study from Charley Wu:* Charley Wu et al., "Similarities and Differences in Spatial and Non-Spatial Cognitive Maps," *PLOS Computational Biology* 16, no. 10 (2020).

153 *Quantum computing's origin story:* the account of Deutsch's insights is from Galchen, "Dream Machine."
154 *prompted in part by his son:* W. Daniel Hillis, "Richard Feynman and the Connection Machine," in *Feynman and Computation*, ed. Anthony Hey (Boca Raton: CRC Press, 2002).
155 *an analysis of all 785,000 articles:* Jian Wang et al., "Bias Against Novelty in Science: A Cautionary Tale for Users of Bibliometric Indicators," *Research Policy* 46, no. 8 (2017).
155 *the last person to know all of mathematics:* In Eric Temple Bell's *Men of Mathematics, Volume Two* (London: Penguin, 1937), the chapter on Poincaré is titled "The Last Universalist."
155 *"domains which are far apart":* Henri Poincaré, "Mathematical Creation," in *The Creative Process: A Symposium*, ed. B. Ghiselin (Berkeley: University of California Press, 1952).
155 *"expanding the adjacent possible":* F. Tria et al., "The Dynamics of Correlated Novelties," *Scientific Reports* 4 (2014).
156 *When the concepts under study are more distant:* Andrey Rzhetsky et al., "Choosing Experiments to Accelerate Collective Discovery," *PNAS* 112, no. 47 (2015).
156 *on a visit to the Van Gogh Museum:* Sachin Waikar, "What Triggers a Career Hot Streak?" *KellogInsight*, October 4, 2021.
156 *Wang is an expert on "hot streaks":* Lu Liu et al., "Hot Streaks in Artistic, Cultural, and Scientific Careers," *Nature* 559 (2018).
157 *"If you look at his production":* quoted in Amanda Morris, "What Was Really the Secret Behind Van Gogh's Success?" *Engineering News (Northwestern)*, September 13, 2021.
158 *Wang's 2021 Nature Communications paper:* Lu Liu et al., "Understanding the Onset of Hot Streaks Across Artistic, Cultural, and Scientific Careers," *Nature Communications* 12 (2021).
158 *the university's mandatory retirement policy:* Carol V. Robinson, "John Fenn (1917–2010)," *Nature* 469 (2011).
159 *"If you just do one or the other":* quoted in Waikar, "What Triggers a Career Hot Streak?"
159 *"Formal Theory of Creativity, Fun, and Intrinsic Motivation":* Jürgen Schmidhuber, *IEEE Transactions on Autonomous Mental Development* 2, no. 3, (2010).
160 *whose much-cited 1991 paper galvanized interest:* James March, "Exploration and Exploration in Organizational Learning," *Organizational Science* 2, no. 1 (1991).
160 *a few backstreets in an obscure subdivision:* the title of my PhD thesis was "Acoustoelectric Interactions in Resonant Tunnelling Structures." It mainly involved sending surface acoustic waves (sort of like a microscopic earthquake) across the surface of a specially designed gallium arsenide chip to see what would happen.
161 *a fifty-year time horizon:* It's hard to get clear information about the mandate

of an organization like the NSA, sometimes referred to as "No Such Agency," because of its all-encompassing secrecy. The fifty-year time horizon is something I was told early in my tenure there, but I don't know if it was an official policy.

161 *Building a working quantum computer*: explanations of how quantum computing works tend to be either simple but wrong or correct but incomprehensible. For an intro that lies somewhere in the middle, try Brian Clegg, *Quantum Computing: The Transformative Technology of the Qubit Revolution* (London, Icon Books, 2021).

162 *"Why sir, there is every probability"*: quoted in William Edward Hartpole Lecky, *Democracy and Liberty, Volume 1* (London: Longmans, Green, & Co., 1899).

162 *"The Usefulness of Useless Knowledge"*: by Abraham Flexner, in *Harper's Magazine* 179, no. 6 (1939).

163 *"As I saw him sitting"*: quoted in Flexner, "The Usefulness of Useless Knowledge."

163 *thirty-five Nobel Laureates*: as of 2024, according to the Institute for Advanced Study's "Mission & History" page at www.ias.edu.

163 *"I dissent from your fundamental proposition"*: quoted in Thomas Neville Bonner, *Iconoclast: Abraham Flexner and a Life in Learning* (Baltimore: Johns Hopkins University Press, 2002).

164 *by around 1980 a transition had occurred*: Paul Forman, "The Primacy of Science in Modernity, of Technology in Postmodernity, and of Ideology in the History of Technology," *History and Technology* 23, no. 1 (2007).

164 *0.8 percent of gross domestic product*: Robbert Dijkgraaf, "Curiosity-Driven Knowledge Is a Vital Form of Infrastructure," *Scientific American*, June 1, 2017.

164 *"No one wants to fund"*: Gina Kolata, "Grant System Leads Cancer Researchers to Play It Safe," *New York Times*, June 27, 2009.

164 *every dollar spent on basic research returns as much as eight dollars*: this is, admittedly, an upper bound. A literature review prepared for the European Commission in 2015 found that "the overall value generated by public research is between three and eight times the initial investment over the entire life cycle of the effects." Luke Georghiou, *Value of Research: Policy Paper by the Research, Innovation, and Science Policy Experts (RISE)* (Luxembourg: Publications Office of the European Union, 2015).

164 *Google announced*: Sundar Pichai, "What Our Quantum Computing Milestone Means," *Google Blog*, October 23, 2019; Frank Arute et al., "Quantum Supremacy Using a Programmable Superconducting Processor," *Nature* 574 (2019); Scott Aaronson, "Why Google's Quantum Supremacy Milestone Matters," *New York Times*, October 30, 2019.

165 *upstart rivals in China and Canada*: Philip Ball, "Physicists in China Challenge Google's 'Quantum Advantage,'" *Nature*, December 3, 2020; Lars Madsen et al., "Quantum Computational Advantage with a Programmable Photonic Processor," *Nature* 606 (2022).

165 *the factors of fifteen*: L. Vandersypen et al., "Experimental Realization of Shor's Quantum Factoring Algorithm Using Nuclear Magnetic Resonance," *Nature* 414 (2001).

165 *"still near the start"*: "A Precarious Milestone for Quantum Computing," *Nature* 574 (2019).

165 *"I never would have conceived"*: Michael Polanyi, "The Potential Theory of Adsorption," *Science* 141, no. 3585 (1963).

CHAPTER 9: THE PROBLEM WITH PASSIVE

169 *Nellie Bly, the New York World's star*: my account of Nellie Bly's voyage is drawn primarily from her book, *Around the World in Seventy-Two Days* (New York: Pictorial Weeklies, 1890), as well as Matthew Goodman's exceptional and definitive account, *Eighty Days: Nellie Bly and Elizabeth Bisland's History-Making Race Around the World* (New York: Ballantine, 2013).

169 *killed an angry reader*: Tim O'Neil, "Looking Back: The Day Post-Dispatch Editor Killed His Adversary in the Newsroom," *St. Louis Post-Dispatch*.

170 *"Superduper neuropsychologists"*: Robert Nozick, *Anarchy, State, and Utopia* (New York: Basic Books, 1974).

171 *"For the vast majority of their members"*: Robert Putnam, *Bowling Alone: The Collapse and Revival of American Community* (New York: Simon & Schuster, 2000).

172 *"Instead of using our physical"*: Mihaly Csikszentmihalyi, *Flow: The Psychology of Optimal Experience* (New York: Harper and Row, 1989).

174 *Thomas Cook, a nineteenth-century businessman*: Edmund Swinglehurst, *Cook's Tours: The Story of Popular Travel* (Poole: Blandford Press, 1982); Stephen Usherwood, "Travel Agents Extraordinary," *History Today* 22, no. 9 (1972). Goodman's *Eighty Days* provides further details about Cook's effect on tourism.

174 *sparked Jules Verne's idea*: the evidence that Verne was inspired by Cook's world tour is circumstantial. Verne himself, in 1898, cited a newspaper advertisement for Cook's tour as his inspiration. But he made other conflicting statements about where the idea had come from. For more, see William Butcher's appendix, "Principal Sources," in the Oxford University Press edition of Verne's book (1995).

175 *"You still have to climb"*: Luke Harding, "Everest's Decline Blamed on Trail of Rich Tourists," *Guardian*, May 17, 2003.

175 *Nepal banned trekking without a guide*: Siobhan Warwicker, "Nepal Imposes Ban on Mountain Trekking Without a Guide," *Guardian*, March 30, 2023.

175 *a 2010 trip to Papua New Guinea*: Alex Hutchinson, "In Papua New Guinea, Trek on a Trail of War," *New York Times*, October 22, 2010.

176 *"The Right to Risk in Wilderness"*: Leo McAvoy and Daniel Dustin, *Journal of Forestry* 79, no. 3 (1981). I wrote about it and interviewed Dustin, for "Call of the Wild," *The Walrus*, June 2013.

177 *"the death of exploration"*: Chris Murphy, "Algorithms Are Making Kids Desperately Unhappy," *New York Times*, July 18, 2023.

178 *"The Google map on the phone in your pocket":* I interviewed Colin Ellard for "Track Record: Why Geotracking Technology Helps Us Find, and Lose, Our Sense of Place," *Canadian Geographic,* May/June 2021.

178 *A 2011 study led by Elizabeth Bonawitz:* "The Double-Edged Sword of Pedagogy: Instruction Limits Spontaneous Exploration and Discovery," *Cognition* 120, no. 3 (2011).

179 *yoked a pair of kittens together:* R. Held and A. Hein, "Movement-Produced Stimulation in the Development of Visually Guided Behavior," *Journal of Comparative and Physiological Psychology* 56 (1963).

179 *a human version of the experiment:* Joel Voss et al., "Hippocampal Brain-Network Coordination During Volitional Exploratory Behavior Enhances Learning," *Nature Neuroscience* 14, no. 1 (2011).

180 *The problem with the new protocol:* Hashem Sadeghiyeh et al., "Lessons from a 'Failed' Replication: The Importance of Taking Action in Exploration," preprint posted November 12, 2018, doi.org/10.31234/osf.io/ue7dx.

181 *Ferdinand Magellan:* Laurence Bergreen, *Over the Edge of the World: Magellan's Terrifying Circumnavigation of the Globe* (New York: William Morrow, 2003).

182 *a magazine feature about a woman with no sense of direction:* "Global Impositioning Systems," *The Walrus,* November 2009.

182 *tagged as "Case one":* Giuseppe Iaria et al., "Developmental Topographical Disorientation: Case One," *Neuropsychologia* 47, no. 1 (2009).

183 *more gray matter in their hippocampus:* Véronique Bohbot et al., "Gray Matter Differences Correlate with Spontaneous Strategies in a Human Virtual Navigation Task," *Journal of Neuroscience* 27, no. 38 (2007).

183 *people who play a lot of video games:* Greg West et al., "Impact of Video Games on Plasticity of the Hippocampus," *Molecular Psychiatry* 23 (2018).

184 *Super Mario 64 actually boosted hippocampus volume:* Greg West et al., "Playing Super Mario 64 Increases Hippocampal Grey Matter in Older Adults," *PLOS ONE* 12, no. 12 (2017).

184 *Schizophrenia patients:* Dennis Velakoulis et al., "Hippocampal Volume in First-Episode Psychoses and Chronic Schizophrenia," *JAMA Psychiatry* 56, no. 2 (1999).

184 *family history of depression:* Michael C. Chen et al., "Decreased Hippocampal Volume in Healthy Girls at Risk of Depression," *JAMA Psychiatry* 67, no. 3 (2010).

184 *the brains of forty combat veterans:* Mark Gilbertson et al., "Smaller Hippocampal Volume Predicts Pathologic Vulnerability to Psychological Trauma," *Nature Neuroscience* 5, no. 11 (2002).

185 *Concorde did it in under thirty-two hours:* "Concorde Lands, Breaking Speed Record," *UPI,* August 16, 1995.

185 *International Space Station:* "Where Is the International Space Station?" *European Space Agency,* www.esa.int.

186 *the company had one more role:* Goodman, *Eighty Days.*

CHAPTER 10: REDISCOVERING PLAY

188 *The game was called "explorers"*: this account of Cody Sheehy's misadventure is primarily based on Emma Marris's wonderful account in the November 2018 issue of *Outside*, "How a 6-Year-Old Survived Being Lost in the Woods." Additional details are from "Lost Child Hikes 18 Miles to Safety," *Associated Press*, April 30, 1986; Dick Cockle, "Missing Child's Miracle Feat Sore Feet Now," *The Oregonian*, April 30, 1986.

189 *Jean-Jacques Rousseau figured*: Paul Bateson and Paul Martin, *Play, Playfulness, Creativity and Innovation* (Cambridge: Cambridge University Press, 2013).

189 *Modern biologists hypothesize*: Lee Alan Dugatkin and Sarina Rodrigues, "Games Animals Play," *Greater Good Magazine*, March 1, 2008.

189 *"all the things we don't have to do"*: Johnson's definition, in *Wonderland: How Play Made the Modern World* (New York: Riverhead, 2016), is riffing off Brian Eno's definition of culture.

189 *attempts to define play*: Peter Gray, "Definitions of Play," *Scholarpedia* 8, no. 7 (2013).

190 *the Daily Mail published*: David Derbyshire, "How Children Lost the Right to Roam in Four Generations," *Daily Mail*, June 15, 2007.

191 *may even end up less anxious*: Anna North, "The Decline of American Playtime—and How to Resurrect It," *Vox*, June 20, 2023.

191 *"Don't worry, doc"*: quoted in Michael Bond, *From Here to There: The Art and Science of Finding and Losing Our Way* (Cambridge: Harvard University Press: 2020).

191 *"We followed them everywhere"*: Edward Cornell and Donald Heth, "Home Range and the Development of Children's Way Finding," *Advances in Child Development and Behavior* 34 (2006).

192 *his 1999 book Lost Person Behavior*: Kenneth Hill (Ottawa: National Search and Rescue Secretariat, 1999).

192 *playtime dropped by 25 percent*: Hillary Burdette and Robert Whitaker, "Resurrecting Free Play in Young Children," *JAMA Pediatrics* 159, no. 1 (2005).

193 *thirteen-month-olds were shown two boxes containing*: Zi L. Sim and Fei Xu, "Infants Preferentially Approach and Explore the Unexpected," *British Journal of Developmental Psychology* 35 (2017).

194 *"Even in the context of relatively straightforward exploratory play"*: Junyi Chu and Laura Schulz, "Not Playing by the Rules: Exploratory Play, Rational Action, and Efficient Search," *Open Mind* 7 (2023).

194 *a 1978 book called The Grasshopper*: Bernard Suits, *The Grasshopper: Games, Life, and Utopia* (Toronto: University of Toronto Press, 1978).

195 *a famous paper titled "Because It Is There"*: George Loewenstein, "Because It Is There: The Challenge of Mountaineering . . . for Utility Theory," *Kyklos* 52, no. 3 (1999).

195 *Marc Malmdorf Andersen points out*: the examples that follow are from his book *Play* (Baltimore: Johns Hopkins University Press, 2022).

196 *"half genius and half buffoon"*: the original 1948 letter is collected in Freeman Dyson, *Maker of Patterns: An Autobiography Through Letters* (New York: Liveright, 2018). His revised appraisal is in Freeman Dyson, *From Eros to Gaia* (New York: Pantheon, 1992).

196 *His memoirs are packed*: Richard Feynman, *"Surely You're Joking, Mr. Feynman!": Adventures of a Curious Character* (New York: W. W. Norton, 1985).

197 *Physics disgusts me a little*: the story of the wobbling place is recounted in Feynman, *Surely You're Joking*.

198 *The Flynn Effect*: Clay Risen, "James R. Flynn, Who Found We Are Getting Smarter, Dies at 86," *New York Times*, January 25, 2021.

198 *what she dubbed the "creativity crisis"*: Kyung Hee Kim, "The Creativity Crisis: The Decrease in Creative Thinking Scores on the Torrance Tests of Creative Thinking," *Creativity Research Journal* 23, no. 4 (2011).

199 *the rate of decline "significantly escalated"*: Kyung Hee Kim, "Creativity Crisis Update: America Follows Asia in Pursuing High Test Scores over Learning," *Roeper Review* 43, no. 1 (2021).

199 *Torrance Tests predict adult creative achievement*: Jonathan Plucker, "Is the Proof in the Pudding? Reanalyses of Torrance's (1958 to present) Longitudinal Data," *Creativity Research Journal* 12, no. 2 (1999).

199 *University of Minnesota researchers published an analysis*: Michael Park et al., "Papers and Patents Are Becoming Less Disruptive over Time," *Nature* 613 (2023).

200 *adolescents came up with more creative responses*: Mark Runco and Shawn Okuda, "Problem Discovery, Divergent Thinking, and the Creative Process," *Journal of Youth and Adolescence* 17, no. 3 (1988).

200 *"I connect technology more to passive play"*: from Kyung Hee Kim's podcast interview with Rob Hopkins, September 20, 2018, www.robhopkins.net.

CHAPTER 11: THE EFFORT PARADOX

203 *"one of the greatest relay races"*: "Columbia Captures 2-Mile Relay Title," *New York Times*, April 25, 1926.

203 *newly installed loudspeaker*: Dave Johnson, "About the Relays," 2008, www.pennrelays.com.

203 *world record of 7:42*: "Georgetown Sets New Relay Record," *New York Times*, April 26, 1925.

203 *"fought off two desperate challenges"*: "Plansky Sets Mark in Penn Carnival," *New York Times*, April 24, 1926.

203 *one second off the respective world records*: the 880-yard record was Ted Meredith's 1:52.2, from 1916. Georgetown's 7:42.0, from the 1925 Penn Relays, was the outdoor 4 x 880-yard record, though they had also run 7:41.6 indoors in 1925. *Progression of World Athletics Records*, ed. Richard Hymans (World Athletics, 2024); "Georgetown Sets World Relay Mark," *New York Times*, March 8, 1925.

204 *"Business, as I have seen it so far"*: quoted in Stephen Larsen and Robin Larsen, *Joseph Campbell: A Fire in the Mind*, (New York: Doubleday, 1991).

204 *"Modern questers for secrets"*: Joseph Campbell's master's thesis, "A Study of the Dolorous Stroke," is included as an appendix in *Romance of the Grail: The Magic and Mystery of Arthurian Myth*, ed. Evans Lansing Smith (Novato: New World Library, 2015).

204 *Boston College's lead at the final hand-off*: "Columbia Captures 2-Mile Relay Title," *New York Times*, April 26, 1926; "Defeat of Hussey in Century and New Records Relay Feature," *Allentown Morning Call*, April 25, 1926; "Yale Sets 880 Yard Mark in Penn Relays," *Chicago Daily Tribune*, April 26, 1926; Joseph Campbell with Bill Moyers, *The Power of Myth* (New York: Doubleday, 1988).

205 *"What I dared to suffer"*: Smith, *Romance of the Grail*.

205 *an admiring Bob Dylan*: this and subsequent stories about Campbell are in Larsen, *Fire in the Mind*.

206 *"I never had the ability"*: *The Hero's Journey: Joseph Campbell on His Life and Work*, ed. Phil Cousineau (New York: Harper & Row, 1990).

207 *"Perhaps these early experiences"*: Larsen, *Fire in the Mind*.

207 *"If anyone would ask me what the peaks were"*: Cousineau, *Hero's Journey*.

207 *pain isn't actually what limits*: Walter Staiano et al., "The Cardinal Exercise Stopper: Muscle Fatigue, Muscle Pain or Perception of Effort?" *Progress in Brain Research* 240 (2018).

207 *"the struggle to continue"*: this phrasing originates as a definition of stamina in Roy Baumeister et al., "The Strength Model of Self-Control," *Current Directions in Psychological Science* 16, no. 6 (2007).

207 *"intensification of either mental or physical activity"*: Michael Inzlicht et al., "The Effort Paradox: Effort is Both Costly and Valued," *Trends in Cognitive Sciences* 22, no. 4 (2018).

208 *law of least effort*: this idea has a long history, but its most influential statement came in Clark Hull, *Principles of Behavior* (New York: Appleton-Century-Crofts, 1943), subsequently codified by Richard Solomon in "The Influence of Work on Behavior," *Psychological Bulletin* 45, no. 1 (1948).

208 *IKEA effect*: Michael Norton et al., "The IKEA Effect: When Labor Leads to Love," *Journal of Consumer Psychology* 22, no. 3 (2012).

208 *effort paradox*: Inzlicht, "Effort Paradox."

208 *"a kind of mystical bliss"*: Cousineau, *Hero's Journey*.

209 *train starlings to fly*: A. Kacelnik and B. Marsh, "Cost Can Increase Preference in Starlings," *Animal Behaviour* 63, no. 2, 2002.

209 *The effect shows up in locusts*: Lorena Pompilio et al., "State-Dependent Learned Valuation Drives Choice in an Invertebrate," *Science* 311, no. 5767 (2006).

209 *a game they called beach bowling*: Joshua Rule et al., "Children Selectively Manipulate Task Difficulty When 'Playing for Fun' vs. 'Trying to Win,'" *PsyArXiv* preprint, dx.doi.org/10.31234/osf.io/q7wh4.

209 *"not just trying to maximize extrinsic reward"*: quoted in Elizabeth Ross, "Why Do Children Play?" *Harvard Graduate School of Education: Usable Knowledge*, September 19, 2023.

NOTES

210 *Meaningfulness of Effort scale:* A. V. Campbell et al., "Meaningfulness of Effort: Deriving Purpose from Really Trying," *PsyArXiv*, June 26, 2022; Michael Inzlicht and Aidan Campbell, "Effort Feels Meaningful," *Trends in Cognitive Sciences* 26, no. 12 (2022).

211 *reward people for choosing the harder option:* Hause Lin et al., "An Experimental Manipulation of the Value of Effort," *Nature Human Behaviour* 8 (2024).

211 *"a bold beginning of uncertain outcome":* Joseph Campbell, *The Masks of God: Creative Mythology* (New York: Viking Penguin, 1968); quoted in Phil Cousineau, "The Soul's High Adventure: Campbell's Comparative Mythology," in *Uses of Comparative Mythology: Essays on the Work of Joseph Campbell*, ed. Kenneth Golden (New York: Routledge, 1992).

211 *digital switching:* Katy Tam and Michael Inzlicht, "Fast-Forward to Boredom: How Switching Behaviour on Digital Media Makes People Bored," *Journal of Experimental Psychology: General*, in press (2024).

212 *research project called Narrative Networks:* the archived program description is at www.darpa.mil/program/narrative-networks.

212 *"to win wars not with weapons":* Mark Finlayson and Steven Corman, "The Military Interest in Narrative," *International Journal of Language Data Processing* 1-2 (2013).

212 *His work on the neuroscience of narratives:* Paul Zak, "Why Inspiring Stories Make Us React: The Neuroscience of Narrative," *Cerebrum* 2 (2015); Jorge Barraza and Paul Zak, "Empathy toward Strangers Triggers Oxytocin Release and Subsequent Generosity," *Annals of the New York Academy of Sciences* 1167 (2009).

213 *this one at Georgia Tech:* M. A. Bezdek et al., "Neural Evidence That Suspense Narrows Attentional Focus," *Neuroscience* 303 (2015).

214 *"'Oh, just government secrets!'":* François Truffaut, *Hitchcock*, revised edition (New York: Simon & Schuster, 1985).

214 *the Holy Grail itself was the ultimate MacGuffin:* Norris Lacy, "Medieval McGuffins: The Arthurian Model," *Arthuriana* 15, no. 4 (2005).

214 *"the last eighty yards":* Cousineau, *Hero's Journey*.

215 *a runner named Frikkie Botha:* Alex Hutchinson, "Comrades Marathon: South Africa," *Canadian Running*, November 15, 2010.

216 *an essay for the Washington Post:* Alex Hutchinson, "Before the Marathon, an Ode to the Losers," *Washington Post*, October 23, 2010. In hindsight, my mother was probably right: I cringe a bit while rereading it.

216 *300 percent effort:* Jim Gerweck, "Semper Fi?" *Running Times*, March 2006.

216 *"One might contend that baseball":* John Rawls, "Two Concepts of Rules," *Philosophical Review* 64, no. 1 (1955).

217 *thirty-six-year-old Andy Sloan:* quoted in Ashley Wu, "The Popularity of Marathons," *New York Times*, April 16, 2024.

218 *"Real fun comes from challenges":* quoted in Sebastian Deterding et al., "Mastering Uncertainty: A Predictive Processing Account of Enjoying Uncertain Success in Video Game Play," *Frontiers in Psychology* 13 (2022).

218 *"positive monotonic curve"*: Chris Crawford, "Design Techniques and Ideals for Computer Games," *Byte* 7, no. 12 (1982).
218 *according to Uppsala University game scholar Ernest Adams:* in "The Designer's Notebook: Difficulty Modes and Dynamic Difficulty Adjustment," *Game Developer*, May 13, 2008.
218 *a class action suit against Electronic Arts:* James Batchelor, "'Dynamic Difficulty' Loot Box Lawsuit Against EA Dropped," *GamesIndustry.biz*, March 4, 2021.
218 *"when the challenges are just balanced"*: quoted in Deterding, "Mastering Uncertainty."
219 *a theory of video games based on the goal of mastering uncertainty:* Deterding, "Mastering Uncertainty."
219 *"follow your blisters"*: Joseph Campbell, *Reflections of the Art of Living* (New York: HarperCollins, 1991), quoted in Scott Allison, "Follow Your Bliss and Heroism," *Encyclopedia of Heroism Studies*, August 24, 2023.

CHAPTER 12: THE FUTURE OF EXPLORING

221 *paddling the entire length of the Again River:* Adam Shoalts, *Alone Against the North: An Expedition into the Unknown* (Toronto: Penguin, 2015).
221 *nine hundred thousand square miles:* Shoalts, *Alone Against the North*.
222 *"Canada's Indiana Jones"*: Jim Rankin, "Explorer Adam Shoalts Heads Back into the Wild Near James Bay—Alone," *Toronto Star*, August 6, 2013.
224 *"when he accidentally canoed over them"*: Kate Harris, "The Future of Exploration," *The Walrus*, June 2018.
224 *"misguided reverence for the lumbering spirit"*: J. R. McConvey, "Review: Adam Shoalts's *Alone Against the North* Feels Powered by Ego as Swollen as a James Bay Thundercloud," *Globe and Mail*, October 9, 2015.
224 *an editorial in Terrae Incognitae:* Lauren Beck, "Firsting in Discovery and Exploration History," 49, no. 2 (2017).
224 *"sterile country"*: Shoalts, *Alone Against the North*.
225 *If there's a path:* quoted in Phil Cousineau, "The Soul's High Adventure: Campbell's Comparative Mythology," in *Uses of Comparative Mythology: Essays on the Work of Joseph Campbell*, ed. Kenneth Golden (New York: Routledge, 1992).
225 *Etienne Brûlé, a Frenchman who came to North America:* details about Brûlé's life are extremely skimpy and sometimes conflicting, primarily from the journals of Samuel de Champlain, *Voyages of Samuel de Champlain 1604–1618*, ed. W. L. Grant (New York: Scribner, 1907), as well as from the French missionaries Gabriel Sagard and Jean de Brébeuf. Further details in French archives were unearthed by Lucien Campeau in the 1970s; and as I write this, new archival findings from historians Éric Brossard and Amandine Lazzarini are emerging (François Bergeron, "Étienne Brûlé N'est Pas Arrivé au Canada Avant 1610," *L'Express de Toronto*, June 15, 2023). Brossard and Lazzarini suggest that, rather than arriving in Canada in 1608, Brûlé didn't get there until 1610. Other dates and details may change as more information emerges; for now I've gone with

the convention assumptions, mostly as outlined in Olga Jurgens, "Brûlé, Étienne," in *Dictionary of Canadian Biography* (Université Laval and University of Toronto, 1966).

225 *the first European to see the Great Lakes:* Brûlé clearly made it to lakes Ontario, Erie, and Huron. He assured the missionary Sagard that there was another lake beyond Huron, and even gave an estimate of its length, which suggests he made it to Lake Superior. There's no evidence that he made it to Lake Michigan; suggestions that he did seem to be based mostly on the idea that he was in the neighborhood en route to Superior.

226 *what is now Pennsylvania:* Brûlé's supposed trip to Chesapeake Bay is based on the story he told Champlain. Not everyone is convinced, though: Donald Kent, "The Myth of Étienne Brûlé," *Pennsylvania History* 43, no. 4 (1976).

226 *learn about their country, see the great lake:* Champlain, *Voyages*.

227 *"untrammeled by man":* Max Greenberg, "A Tribute to Howard Zahniser, Unsung Architect of the Wilderness Act," *The Wilderness Society*, February 25, 2016.

227 *"If nature dies because we enter it":* William Cronon, "The Trouble with Wilderness: Or, Getting Back to the Wrong Nature," *Environmental History* 1, no. 1 (1996).

227 *Roderick Nash: Wilderness and the American Mind* (New Haven: Yale University Press, 1967).

228 *"Thunder and rain prevailed": The Journals and Letters of Sir Alexander Mackenzie*, ed. W. Kaye Lamb (Cambridge: Cambridge University Press, 1970).

229 *a 2,500-mile canoe journey:* Adam Shoalts, *Beyond the Trees: A Journey Alone Across Canada's Arctic* (Toronto: Allen Lane, 2019).

229 *2,000 miles from his home:* Adam Shoalts, *Where the Falcon Flies: A 3,400 Kilometre Odyssey from My Doorstep to the Arctic* (Toronto: Allen Lane, 2023).

229 *Such game-playing, in Suits's view:* Bernard Suits, *The Grasshopper: Games, Life, and Utopia* (Toronto: University of Toronto Press, 1978).

229 *"occupational methadone":* quoted in Christopher Yorke, "'The Alexandrian Condition': Suits on Boredom, Death, and Utopian Games," *Sport, Ethics and Philosophy* 13, no. 3–4 (2019).

229 *"Suits exaggerates":* Thomas Hurka, "Introduction," in the 2005 Broadview Press edition of Suits, *The Grasshopper*.

231 *"he came upon a little footpath":* Champlain, *Voyages*.

232 *he was killed and eaten:* the idea that Brûlé was eaten comes from the missionary Gabriel Sagard, *Histoire du Canada et Voyages que les Frères Mineurs Recollects y Ont Faicts Pour la Conversion des Infidelles* (Paris: Claude Sonnius, 1836). Whether it's true or not will never be known, but Sagard knew Brûlé well and lived among the Huron-Wendat like Brûlé, and his report is unambiguous.

232 *the ghost of Brûlé's sister:* from *Les Relations des Jésuites*, quoted in Stéphanie St-Pierre, Du Traître au Héros: Étude Historiographique et Critique d'Étienne Brûlé de 1619 à Nos Jours, master's thesis, Laurentian University (2006).

233 *John Cabot was the first post-Viking European:* the historian Alwyn Ruddock

famously claimed to have evidence that Cabot actually made it back to England in 1500, but she ordered all her papers to be destroyed after her death in 2005 and no one has been able to substantiate her claims. "The Discovery of America: The Revolutionary Claims of a Dead Historian," *University of Bristol*, press release, April 4, 2007.

233 *"When there are no more worlds"*: Suits, *The Grasshopper*; see also Yorke, "The Alexandrian Condition."

233 *modern society is "hyperdopaminergic"*: Fred Previc, *The Dopaminergic Mind in Human Evolution and History* (Cambridge: Cambridge University Press, 2009).

234 *dopamine fasting*: Nellie Bowles, "How to Feel Nothing Now, in Order to Feel More Later," *New York Times*, November 7, 2019.

234 *Brûlé was rediscovered by historians*: St-Pierre, *Du Traître au Héros*.

235 *another historian, Éric Brossard, unearthed*: Danièle Caloz, "Étienne Brûlé: A Wealthy Parisian Trader?" *The Canadian Encyclopedia*, March 16, 2015.

238 *links between exploratory behavior and the microbes*: Monica McNamara et al., "Oral Antibiotics Reduce Voluntary Exercise Behavior in Athletic Mice," *Behavioural Processes* 199 (2022).

239 *the "oldest-old" residents*: Deborah Grady et al., "DRD4 Genotype Predicts Longevity in Mouse and Human," *Journal of Neuroscience* 33, no. 1 (2013).

EPILOGUE

241 *"the active, difficult, wandering, dissipated life"*: Denis Diderot, "Supplement to Bougainville's 'Voyage,'" in *Rameau's Nephew and Other Works*, (Indianapolis: Hackett, 2001).

241 *"continents to invade"*: Denis Diderot, in *A History of the Two Indies: A Translated Selection of Writings from Raynal's Histoire philosophique et politique des établissement des Européens dans les Deux Indes*, ed. Peter Jimack (Routledge: London, 2006). Several authors contributed to Guillaume-Thomas Raynal's book; historians attribute the passages quoted here to Diderot.

241 *"ambition, misery, curiosity"*: this is also from Raynal's *Histoire*, quoted in Anthony Pagden, "The Effacement of Difference: Colonialism and the Origins of Nationalism in Diderot and Herder," in *After Colonialism*, ed. Gyan Prakash (Princeton: Princeton University Press, 1995).

242 *motivational pluralism*: Paul Bloom, *The Sweet Spot* (New York: Ecco, 2021).

243 *"interested in the uncertainty reduction"*: quoted in Christie Aschwanden, "Uncertainty Is Science's Superpower. Make It Yours, Too," April 3, 2024, ep. 2 of the Scientific American podcast *Uncertain*.

244 *"The allure of the great potential"*: Michael Distefano, "Sealed Case of 1979-80 OPC Hockey Cards Containing Gretzky Rookie Cards Sells for $3.72M." *Hockey News*, February 25, 2024.

244 *"forty freedoms"*: Aldo Leopold, *A Sand County Almanac* (New York: Oxford University Press, 1949).

INDEX

absolute exploration, 228–29
accidental drift, 26
Account of the Polynesian Race, Its Origin and Migrations, An (Fornander), 23–24, 25
Adams, Ernest, 218
"adaptive flexibility," 34–35
addiction, 223
   dopamine and, 40–41
advertising, 63
aesthetics, 159–60, 210
African origins of modern humans, 13, 20–21, 49–50
Again River, 221–22, 223–25, 229, 230
agriculture, 50, 51, 122, 227
AIBO, 67–69
Alaska, 21, 96–97, 105, 176
Alexander the Great, 233
"Alexandrian condition," 233
Algonquin Park, 2
Algorithms to Live By (Christian and Griffiths), 121
allocentric cognitive maps, 143–44
Almgren, Andreas, 94
Alone Against the North (Shoalts), 221
Alzheimer's disease, 183
ambiguity, 100–103, 108–9
ambiguity aversion, 101–2, 103, 107–10
   extreme, 102, 127–29
American Cancer Society, 164
Amundsen, Roald, 233
Ancient Voyagers in the Pacific (Sharp), 21–22, 24–28
Andersen, Marc Malmdorf, 123–24, 189, 219, 220
   music and, 124, 125, 195, 201–2
   theory of play, 64, 123–24, 126, 159, 189, 195, 200, 201–2, 206, 210
animal studies, 46–48, 83. *See also specific animals*
Antarctic, 106
anxiety, 93, 136, 238

Apple iPhones, 183
applied research, 160–62
Arctic, 111, 112, 176, 177, 229, 233
Arctic Circle, 98
Arctic Ocean, 103–4, 229
Ariaal tribe, 50–51
Aristotle, 230
Armstrong, Louis, 123
Army Ranger Battalion (Sweden), 75
Around the World in Eighty Days (Verne), 169, 174
around-the-world trip of Nellie Bly, 169–70, 173–74, 180–81, 185–87
Arthurian legends, 204–5, 214
artificial curiosity, 69
artificial intelligence (AI), 69, 81, 122, 159–60, 223
artists and hot streaks, 156–59
Aschwanden, Christie, 243
Astrosmash (video game), 218
Atari, 218
Athabasca, Lake, 96, 97, 111
Attenborough, David, 177
attention deficit hyperactivity disorder (ADHD), 38–39, 51
Augusta Victoria, SS, 180
Australia
   cane toads, 47
   early European discovery and exploration, 135–38
   early human migrations, 21, 29–30
   first crossings of, 130–31, 137–38, 139–44, *143*, 146–48
   Ningaloo Reef, 4
Australian National University, 30
Australian Overland Telegraph Line, 137–38, 147
Awakenings (Sacks), 42, 183

babies (infants), 39, 66, 68, 83, 122, 159, 193–94

backpacking, 4
  Long Range Traverse, 1–3, 4–5, 9–10, 11
  Willmore Wilderness Park, 3–4
Bad Singer (Falconer), 125–26
Ballard, Dana, 56
Bamberger, Louis, 162
bandit problems, 79–83, 85–86, 88–90
bandy, 75
Banks, Joseph, 24
Bantu culture, 50
Bar, Moshe, 70–72, 238
Beard, Charles, 163
"Because It Is There" (Loewenstein), 195
Beck, Lauren, 224
bees, 76–77
behavioral modernity, 20, 29–30, 51, 144–45
behaviorism, 133–34
Bella Coola River, 112
Bell Labs, 150
Bellman, Richard, 119
Berlyne, Daniel, 19, 107
Berridge, Kent, 42, 43
big data, 155
"Big Five" personality traits, 210–11
bipolar disorder, 55
Bird (movie), 123, 124, 125
birds, 17, 29, 46, 47, 141
Bisland, Elizabeth, 180–81, 185–87
"bits," 150
Blackwell's Island Insane Asylum, 170
blank spots, 85, 103, 110, 187, 221, 222, 245, 246
blicket experiments, 116–17, 118
blockbuster movie sequels, 121–22
Bloom, Paul, 71, 242
Bly, Nellie, 169–71, 173–74, 180–81, 185–87, 243
Bohbot, Véronique, 145, 148, 182–85, 243
Bonawitz, Elizabeth, 109–10, 178–79, 209–10
boredom, 62, 66, 159, 195, 211, 229
Boston College, 203–4, 206–7
Boston Marathon, 217
Botha, Frikkie, 215
Bothnia, SS, 186
bottom-up processing, 55–56, 71
Bourdelle, Antoine, 206
Bowling Alone (Putnam), 171–72
Boykov, Luben, 13
Bradt, Tyler, 99

Brahé, William, 130–31, 146
brain
  cognitive maps. *See* cognitive maps
  cognitive shortcuts, 152–53
  developmental topographical disorientation, 182–85
  development of, 120
  dopamine and, 9, 38, 39–44, 69, 82–83, 233–34, 239
  evolution of, 116–17, 145
  experience machine, 170–71
  heuristics, 81
  narratives and, 212–13
  perception of time and, 54
  predictive processing. *See* predictive processing
brain-derived neurotrophic factor (BDNF), 70
brain imaging, 57, 102, 179, 182, 184, 212–13
brain plasticity, 70
brainstorming, 71–72, 165
Brasher, Chris, 11
Brébeuf, Jean de, 232
British Geological Survey, 28
British Isles, 49–50
Brixel, Richard, 13
Brossard, Éric, 235
Browning, Robert, 99
Bruce Peninsula, 2
Brûlé, Étienne, 225–28, 231–33, 234–35, 245
Bryn Mawr College, 162
Bryson, Bill, 2
Burke, Robert O'Hara, 130–31, 137–38, 139–44, *143*, 146–48, 156
Bush, George W., 161
Bush, Robert, 79–80

Cabot, John, 233
call to adventure, 205, 226–27
Campbell, Aidan, 210
Campbell, Joseph, 10, 203–7, 208, 211–15, 219, 226–27
Campeau, Lucien, 234–35
Canadian Running, 5
cane toads, 47, 49
Carlsson, Arvid, 41–42
Carnegie Mellon University, 63
Casasanto, Daniel, 152
caudate nucleus, 145, 182–85
Cavendish Laboratory, 149–50

challenge, 209-10
  optimal, 66, 217-18, 219
  Wundt Curve, 65, 65-67
Champlain, Samuel de, 225-26, 231-32, 234
Chase, Chevy, 174
Chen, Chuansheng, 38-40, 43, 45, 48, 50, 239
chickens, 116
children (childhood)
  exploration, 115-19, 122-23, 126, 178-79
  explore-exploit dilemmas, 109-10, 116-17
  "home ranges," 190-91, 192-93
  play, 188-96, 200-201, 209-10
  wandering, 188-89, 190-93
chimpanzees, 116
Chown, Jillian, 159
Christian, Brian, 121
chronometers, 96, 97-98
Chu, Junyi, 194-95, 210
Chung, Joanna, 210
citation counts, 155, 156, 160
Ciudad Guayana, 132
Clapham, Vic, 215
Claremont Graduate University, 212
Clark, Andy, 55, 57, 64
Clark University, 132-33
classical conditioning, 209
Coast Mountains, 112
Cockerill, Jack, 169-70
cockroaches, 136
cognitive dissonance, 208-9
cognitive distances, 156
cognitive maps (mapping), 131-35, 151-53
  behavioral modernity and, 144-45
  developmental topographical disorientation, 182-85
  first crossings of Australia, 130-31, 139-44, *143*, 146-48
  use of term, 132
"Cognitive Maps in Rats and Men" (Tolman), 134
Coiffier, Alizon, 235
coin tosses, 100, 109
Colbourne, Clayton, 12-13
Cold War, 100, 163-64
collective unconscious, 205, 212
College of William & Mary, 198
Columbia University, 6, 58, 88, 203-4, 206-7

Columbus, Christopher, 18-19, 85, 113, 224, 228, 233
"combinatorial novelty," 155
Commonwealth Games, 4
commuters, 80-90
Company of Adventurers to Canada, 232
complexity and Wundt Curve, 65, 65-67
computational complexity, 154
Comrades Marathon, 215-17
conceptual exploration, 159-60
Concorde, 185
conscientiousness, 210-11
Continental Divide, 111
contrast theory, 209
Cookie Clicker (video game), 218-19
Cook Inlet, 105
Cook, James, 24, 25-26, 31-32, 96-97, 105, 241
Cook, Thomas, 174-75. *See also* Thomas Cook & Son
Cooper Creek, 130-31, 139, 140-41, 144, 146-47, 147-48
Cornell, Ed, 191-92, 193
Cornell University, 197
Cosmopolitan, The, 180
Covid-19 pandemic, 5, 160, 245-46
Crawford, Chris, 218
creativity, 198-200, 233
  artists and hot streaks, 156-59
"creativity crisis," 198-200
creativity tests, 190, 198-99
Creek, James, 111
Crete, 29, 34
Cromer Forest Bed, 28
Cronon, William, 227
cryptography, 150-51
Csikszentmihalyi, Mihaly, 172-73, 218
cultural knowledge, 115, 118-19
curiosity, 63, 237
  in animals, 46, 48, 133
  in research, 162, 164, 196

Daley, Francis, 206-7
dandy brushes, 139-40
Dark Room problem, 58-62
Darwin, Charles, 85, 189
Davies, Robertson, 125
daydreaming, 71
decision-making, 8, 79-80
Decker, George, 12
decoherence, 161
Defense Advanced Research Projects Agency (DARPA), 212-13

INDEX

Deliveroo, 86–87
Denisovans, 20
depression, 55, 58, 112, 183, 184
Descent of Man, The (Darwin), 189
Deterding, Sebastian, 219
Deutsch, David, 150, 151, 153–54
developmental topographical disorientation (DTD), 182–85
Diamantina River, 142
Dickens, Charles, 115
Diderot, Denis, 241–42
diffusion equation, 18, 29
"digestible errors," 64–65, 72
digital switching, 211
Dig Tree, 146, 148
Diller, Barry, 121
direction sampling, 192
"disruptiveness score," 199
DiStefano, Michael, 244
DNA (deoxyribonucleic acid), 20–21, 39, 40, 43, 46, 49
Dobbs, David, 49
doldrums, 17, 31–33
Don Quixote (Cervantes), 7
Doom (video game), 60
"doorway effect," 135
dopamine, 9, 38, 39–44, 69, 82–83, 233–34, 239
dopamine fasting, 234
Dopamine Nation (Lembke), 40
Dopaminergic Mind in Human Evolution and History, The (Previc), 233–34
DRD4 (dopamine receptor D4), 39–40, 43, 45–48, 50–51, 66, 239
dreaming, 71
Dudchenko, Paul, 133
Durand, Jorge, 44–46
Dustin, Daniel, 176–77
Dutch East India Company, 135–36
Dylan, Bob, 205
dynamic difficulty adjustment (DDA), 218–19
Dyson, Freeman, 196

early human migration. *See* Great Human Expansion
Easter Island, 22, 26
Eastwood, Clint, 123
effort, 207–11
effort paradox, 208–9, 219, 230
egocentric cognitive maps, 143–44
Ehrlich, Paul, 162–63, 164

Eichenbaum, Howard, 153
Einstein, Albert, 162, 233
Eisenberg, Dan, 50–51
Elcano, Juan Sebastián, 182
electrons, 161, 197
electrospray ionization, 158
elite athletes, 75–76
  Van der Poel's story, 75–76, 77–78, 83–85, 94–95
Ellard, Colin, 178
Ellsberg, Daniel, 100–103, 107–10
Ellsberg paradox, 101–2
embrace the struggle, 238
emigration, 36–37, 44–46
Endure (Hutchinson), 5–6
English Channel, 29
entropy, 58, 158
Erie, Lake, 226
Erikson, Leif, 11–13
Etch A Sketch, 135
Étienne Brûlé Park (Toronto), 226
Etruria, RMS, 186
European Vacation (movie), 174
evolution, 7, 18, 20–21, 50, 59, 116–17, 136
exercise, 70, 76, 84, 120, 238–39
experience machine, 170–71
exploitation, 7
exploitatory state of mind, 71–72
exploit dilemmas. *See* explore-exploit dilemmas
explorare, 10
"Exploration and Exploitation in Organizational Learning." (March), 7–8, 160, 162, 164
exploration, use of term, 7, 10–11
exploratory behavior, 19, 43, 46–47, 68, 89, 121
exploratory play, 193–94, 202, 205
exploratory state of mind, 71–72
explore-exploit dilemmas, 6–10, 75–95, 119, 151–51
  explore *then* exploit, 236–37
  March on, 7–8, 76, 160, 162, 164
  multi-armed bandits, 79–83, 85–87, 118, 119, 236–37
  optimal stopping problems, 91–94
  in preschoolers, 109–10
  use of term, 7, 76–77
  Van der Poel's story, 75–76, 77–78, 83–85, 94–95
"explorer's gene," 40, 51
explorers' journals, 85

exploring algorithms, 7–8, 81, 86, 87–88
extreme ambiguity aversion, 102, 127–29

failure, 10, 14, 205, 211, 215–16, 217, 219
fairy terns, 17
Falconer, Tim, 125–26
Faraday, Michael, 162, 164
"feeling of grip," 63–64
Fee, Michale, 83
Fenn, John, 158
Ferguson, Thomas, 93
Fernández-Armesto, Felipe, 20, 21, 22
Ferranti Atlas, 26
Feynman, Richard, 150, 151, 154, 196–98, 201
Fiancé Problem, 91–92, 93–94
Fick, Adolf, 18
Fiji, 23, 26
"filter bubbles," 178
Finney, Ben, 21, 23–24, 26–28, 30, 31–34
Finney, Ruth, 28
fire, 114–15
firsting, 224, 225, 228
first-person shooter video games, 60, 184
Firth, Raymond, 24
fish, 59, 103
Fleming, Alexander, 195–96
Flexner, Abraham, 162–63
Flood, Merrill, 91–92, 94
Flood, Sue, 91–92, 94
Flores Island, 20, 29
Flores man (Hobbit), 20
Florida State University, 30
flow, 172–73, 218
Flynn effect, 198–99
Flynn, James, 198
folk wisdom, 192
"follow your bliss," 206, 219
"Formal Theory of Creativity, Fun, and Intrinsic Motivation" (Schmidhuber), 159–60
Forman, Paul, 164
Fornander, Abraham, 23–24, 25, 34
Fort Chipewyan, 96, 98, 105, 107, 111, 112
Fort, Joaquim, 18, 29
Fourteenth Street Bridge, 216
Franklin, John, 113
Fraser River, 111
Fraser, Simon, 111–12
Freed, Peter, 58, 60

free energy principle, 57–58, 59–61, 63–64, 72, 99
French-Canadian settlers, 48–49
"frequency dependent selection," 50
Friendly Village, 112
Friston, Karl, 57–58, 59
frogs, 46
Frontier Thesis, 227
Fuld, Caroline, 162
fun, 159–60, 189, 194, 209, 218, 237
future of exploration, 221–39

Gabor patches, 153
Gaffney, Dylan, 34
Gage, Timothy, 45–46
Garcia, Jerry, 205
Generation Z, 122
gene surfing, 48–49
genetics, 39–40, 45–52, 55, 239
genomics, 39
geographical exploration, 13–14, 17–19, 135–36, 181–82, 224, 225–28, 231–33. *See also specific explorers*
  first crossings of Australia, 130–31, 139–44, *143*, 146–48
  Mackenzie's expeditions, 96–98, 99, 103–7, *104*, 111–13
  Polynesia, 22–26, 31–34
Geological Survey of Canada, 221
Georgia Institute of Technology (Georgia Tech), 8, 67, 107, 116, 213
Giron, Anna, 118
Gittins index, 80–81, 82
Gittins, John, 76, 80–81, 82, 119
GKN Aerospace, 95
Gladstone, William, 162
Globe and Mail, 5
Gödel, Kurt, 162
"Goldilocks effect," 66
Goodman, Matthew, 181
Google, 164–65, 178
Google Scholar, 57
Gopnik, Alison, 116–19, 122–23, 126
GPS (global positioning system), 11, 69, 152, 176, 182, 183, 185, 242–43
Grant, Cary, 213–14
Grasshopper, The (Suits), 194–95, 229–30
Grateful Dead, 205
Gray, Charles, 130–31
Great Barrier Reef, 4
Great Dividing Range, 137
Great Exhibition of 1851, 174

Great Human Expansion, 13, 20–21,
    29–30, 34–35, 43–44, 49–52, 55
  novelty-seeking genes and, 39–40
  Polynesians, 21–38
Great Inland Sea, 137
Great Lakes, 225–27
Great Slave Lake, 97–98, 105
Greenland, 11–12, 21
Gregory, Augustus, 147
Gretzky, Wayne, 244
"grid cells," 135
Griffiths, Tom, 121
"grit," 210–11
Gros Morne National Park, 1–3, 4–5,
    9–10, 11
Guggenheim, Peggy, 157–58
Gulf of Carpentaria, 142
Gulf of Mexico fishing, 89
Gulf of Tonkin incident, 102–3
gut microbiome, 238–39

habits, 8–9, 70–71, 90–91
hallar, 224
Halley's Comet, 200
Hampton Court Palace, 133
Harris, Kate, 224
Hartmann von Aue, 204–5
Harvard University, 49, 79, 91, 100, 102,
    170, 171, 178, 184, 209–10, 212
Hawaii, 22, 23, 26, 27, 30, 31, 32, 206
Heisenberg, Werner, 154
Helmholtz, Hermann von, 56
Hendricks, Mike, 55, 69
hero's journey, 204–6, 213, 219
Hero with a Thousand Faces, The
    (Campbell), 206
Herrick Creek, 111
Herrick, John, 102–3
Herz, Noa, 71–72
Heth, Donald, 191–92, 193
heuristics, 81
Heyerdahl, Thor, 33
"high anxiety sensitivity," 136
Hillary, Edmund, 233
Hill, Kenneth, 192
Hindu Kush, 23
hippocampus, 134–35, 136–37, 145, 152,
    153, 179, 182–85, 243
Hitchcock, Alfred, 213–14
hobbits, 20
Hokule'a, 27, 28, 31–34, 38
Holmes, Tommy, 28
Holy Grail, 204, 214, 230

"home ranges," 190–91, 192–93
Homo erectus, 20
Homo sapiens, 20–21, 34–35
Hong Kong, 181, 185
Honolulu Marathon, 217
Horizon Task, 108–9, 119–20, 127,
    179–80
horses, 116
hot streaks, 156–59
Howe, Kerry, 25
Hsee, Christopher, 62, 63
Hudson Bay Lowlands, 221, 224
Hull, Clark, 208
human endurance, 5–6, 207, 215–16
human height, 50
human migration. *See* migration
Humber River, 226, 227, 234, 245–46
hunter-gatherers, 50, 51, 114–15, 116,
    144–45, 234
Hurka, Thomas, 230
Huron, Lake, 226
Huron-Wendat, 227–28, 231, 232
Hurricane Hazel, 226
hydroxytyramine. *See* dopamine
hyperactivity, 38–39, 51
Hyperborean Sea, 98
hyperdopaminergic, 233–34

Iaria, Giuseppe, 145, 182–83
IBM, 164
  Deep Blue, 67
Icelandic Sagas, 11–12
ideas, mapping, 149–65
  artists and hot streaks, 156–59
idle games, 218–19
IKEA effect, 208
Immersion Neuroscience, 213
Industrial Revolution, 115, 122
information bonuses, 109–10
information gain, 109, 110, 193
information theory, 56
Ingstad, Anne Stine, 12
Ingstad, Helge, 11–12
Institute for Advanced Study, 162–63
intention, 30
"intent to explore," 26
International Arthurian Society, 214
International Monetary Fund, 90
International Space Station, 185
intuition, 57–58, 81–82, 126
Inzlicht, Michael, 208–9, 210–11, 214,
    219
IQ tests, 198–99

INDEX

Iraq WMDs, 99, 103
Iroquois, 227–28, 231

Jackson, Peter, 158
Jacobs, Jane, 136
James Cook University, 139–40
Janszoon, Willem, 135–36
Java, 137
jazz, 123, 124, 125
Jefferson, Thomas, 112–13
job-hiring decisions, 92
job skills, 122
Johaug, Therese, 84
Johnson-Schwartz, J. S., 19
Johnson, Stephen, 189
Jones, Pei Te Hurinui, 25
Jung, Carl, 85, 212
Jupiter, 105

Kamehameha III, 23
Kane, Herb, 28
Kant, Immanuel, 189
Kapahulehua, Kawika, 33
Kaplan, Frédéric, 67–69
Kasparov, Garry, 67
Keats, John, 99
Kepler, Johannes, 93–94
Keynes, John Maynard, 99–101
Kidd, Celeste, 66
Kim, Kyung Hee, 198–201
King, John, 131, 146
King's College London, 60
King's Trail, 77
Kipchoge, Eliud, 217
Kirke, David, 232
Kiverstein, Julian, 63–64
Knight, Frank, 99–100
"Knightian uncertainty," 100
known knowns, 99
known unknowns, 99
Kokoda Track, 175–76
Kon-Tiki expedition, 33
Koppisch, Walter, 206
Koster, Raph, 218
Krakauer, David, 243–44
Krishnamurti, Jiddu, 206

lactate threshold, 54
Lacy, Norris, 214
Lai, Tze Leung, 88
Landauer, Rolf, 150–51
landmarks and ruins, 11–13
language and maps, 152

L'Anse aux Meadows, 12–13
Lapidow, Elizabeth, 109–10
Larson, Jeremy, 126
Lashley, Karl, 134
law of least effort, 208
L-dopa (levodopa), 42
learned industriousness, 209
"learning traps," 117
Le Champagne, SS, 186
Le Havre, 186
Leichhardt, Ludwig, 138
Leifsbudir, 12
Leisure World, 239
Leopold, Aldo, 244–45
Leppard, Thomas, 30
Lewis and Clark Expedition, 113
Lewis, David, 30–32, 33
Liquori, Marty, 54
Loewenstein, George, 63, 106, 195
London taxi drivers, 145, 183
London Tube strike of 2014, 89–90, 91
Long Range Traverse, 1–3, 4–5, 9–10, 11
Lord of the Rings (movies), 158
Los Alamos National Laboratory, 196, 197
Lost Person Behavior (Hill), 192
Lucas, George, 205–6
ludic behavior, 19, 107
Luomala, Katherine, 21

McAvoy, Leo, 176–77
McCarthy, Cormac, 243
McGill University, 145, 182
MacGregor River, 111
MacGuffins, 213–15, 230, 243
machine learning, 60–61, 81
Mackenzie, Alexander, 96–98, 99, 103–7, *104*, 111–13, 140, 165, 228, 243
Mackenzie River, 103–4, 106–7, 113
McMaster University, 70
McNamara, Robert, 103
Madagascar, 34
Maddox, USS, 102
Mad Men (TV show), 92
Magellan, Ferdinand, 22, 181–82, 187
Mallory, George, 19, 208
Manhattan Project, 163, 197
Man Who Knew Too Much, The (movie), 213–14
March, James, 7–8, 76, 160, 162, 164
Marine Corps Marathon, 216, 217
Marquesas Islands, 22
Marris, Emma, 202

Mars, 19, 52, 222
Marx, Karl, 230
Mason, James, 213–14
Massachusetts Institute of Technology (MIT), 56, 83, 150, 194
Massey, Douglas, 44–46
Matrix, the, 171
Matthews, Luke, 50, 51–52
Mawona, Dudley, 215
Max Planck Institute for Biological Cybernetics, 86
Maxwell, James Clerk, 149
Ma, Yo-Yo, 177
maze studies, 109, 132–35, 138–39, 140–41
meaningful exploration, 10–11, 14
Meaningfulness of Effort scale, 210–11
Meeting of Two Worlds, The (Boykov and Brixel), 13
Mendaña de Neira, Álvaro de, 22
Menindee, 140, 146
Mexican emigration, 36–37, 38, 44–46
Mexican Migration Project (MMP), 36, 44–46
mice, 136, 138, 139, 140–41, 238–39
Michigan, Lake, 225–26
migration. *See also* Great Human Expansion
 DRD4 and, 39–40, 45–47, 49–52, 239
 of Mexicans, 36–37, 38, 44–46, 72
 network theory of, 44–45
 Out of Africa hypothesis, 13, 20–21, 49–50
 of Polynesians, 21–34, 37–38
Milford Track, 4
Military Academy, U.S. (West Point), 55, 69
military decision-making, 99–100, 102–3
Miller, Mark, 59, 60, 63–64, 67, 69, 72, 223
mindfulness, 71
Moana (movie), 24
Moby Dick (Melville), 24
modern tourism, 174–76
Molecule of More, The (Lieberman and Long), 40
Montagu, Katharine, 41
Moore's Law, 199
Moose Cree First Nation, 224
Moran, Rosalyn, 60–61
Morris, Richard, 138
Morris water maze, 138–39, 140–41
Moser, Edvard, 135
Moser, May-Britt, 135

Mosteller, Frederick, 79–80
motivational pluralism, 242
"motor babbling," 68
motor control and dopamine, 42
Mount Everest, 19, 77, 175, 187, 208, 229, 233
Mount Hopeless, 146, 147–48
Mozart, Wolfgang Amadeus, 195, 201–2
Mtshali, Robert, 215
Multi-Armed Bandit Allocation Indices (Gittins), 80–81
multi-armed bandits, 79–83, 85–87, 98, 118, 119, 179, 236–37
Murgatroyd, Sarah, 139
Murphy, Chris, 177–78
musical exploration, 123–27

nanoelectromechanics, 161–62
Napierala, Jeffrey, 45–46
Napoleon Bonaparte, 233
Narrative Networks, 212–13
Nash, Roderick, 227–28
National Geographic, 21, 175–76
National Security Agency (NSA), 6, 151, 160–62, 165
Native Americans, 40, 111–12, 226, 227–28, 231, 232
natural selection, 50, 59
Nature (journal), 165
Nature Communications, 158
Neanderthals, 18, 20–21, 34, 144–45
Nepal, 175
nervous system, 55–56
Nestabeck, 97
network theory of migration, 44–45
neurons, 41–42, 43, 83, 120, 134–35, 136, 152
neuroscience. *See* brain
neurotransmitters, 41–42, 83
Newfoundland, 1–3, 4–5, 9–10, 11–14, 242–43
New Guinea, 136, 175–76
Newport, Rhode Island, 11–12
Newtonian physics, 154
Newton, Isaac, 149, 233
New York Times, 4, 5, 19, 115, 164, 203, 207, 222, 234
New York World, 169–71, 173–74, 187
New Zealand, 4, 21, 23, 25, 26
Niagara Falls, 99
Nietzsche, Friedrich, 230
Nils van der Poel: Genius or Fool? (documentary), 78

Ningaloo Reef, 4
Nixon, Richard, 100
norepinephrine, 83
no-rescue wilderness zones, 176–77
Norgay, Tenzing, 175
North American Aerospace Defense Command, 102
North by Northwest (movie), 213–14
North Pole, 20, 187, 233
Northwestern University, 156, 159
Northwest Passage, 113, 224, 233
novelty (novelty-seeking)
   DRD4 and, 39–40, 46, 47, 50–51
   embrace the struggle, 238
   Wundt Curve, 65, 65–67
"Now's the Time" (song), 125
Nozick, Robert, 170–71

"occupational methadone," 229
Ogilvy, David, 63
O'Keefe, John, 134–35, 136–37, 145
Olympic Games (Beijing; 2022), 77, 78, 84, 94–95
Olympic Games (Pyeongchang; 2018), 75
Olympics training, 6, 77–78, 79, 84–85
Ontario, Lake, 226
O-Pee-Chee Company, 244
optimal challenge, 66, 217–18, 219
optimal stopping problems, 91–94
ordering food, 80, 86–88
orienteering, 53–55, 59, 69–70, 72
Origin of the British Flora, The (Reid), 29
Oudeyer, Pierre-Yves, 68–69
Out of Africa hypothesis, 13, 20–21, 49–50
Outside (magazine), 5, 202
Owen Stanley Mountain Range, 175–76
Oxford flu, 149, 150–51, 160
oxytocin, 212–13

Pääbo, Svante, 20–21
Pacific Ocean, 112, 181–82
package tours, 174
Pack River, 111
Palomar Mountain, 192
Palouse Falls, 99
Papeete, 33
Papua New Guinea, 175–76
Paramount Pictures, 121
Parker, Charlie, 123, 124, 125
Park, Heejung, 115–16, 122
Parkinson's disease, 42
Parks Canada, 2-2, 11, 12, 242

passive exploration, 170–80, 182
   developmental topographical disorientation, 182–85
patents, 155, 156, 158, 199
Pavlov, Ivan, 209
Peace River, 111
penicillin, 195–96
Penn Relays (1926), 203–4, 206–7, 208
Pennsylvania State University, 214
Pentagon Papers, 100, 103
Philippines, 23
Phoenix, David, 139–40, 147–48
physics, 6, 149–50, 196–97
Piailug, Pius "Mau," 17–18, 31–32, 33
Pichler, Wolfgang, 78, 95
Pinhasi, Ron, 49
Pitchfork (magazine), 126
Pitt, Benjamin, 152
Pittsburgh Dispatch, 173
place cells, 134–35, 136–37
planning horizon. *See* time horizon
Plato, 188–89
play, 188–202, 237
   Andersen's theory of, 64–65, 123–24, 126, 159, 189–90, 195, 200, 201–2, 206, 210
   childhood, 188–96, 200–201, 209–10
   link between exploring and, 189–90
   "ludic behavior" as, 19
   purpose of, 188–90
   Suits on, 194–95, 229–30, 233
   video games, 172–73, 183–84
pleasure, 171, 219
   dopamine and, 40–43
PLOS One, 70
Poincaré, Henri, 155
Polanyi, Michael, 165
Pollock, Jackson, 157–58
Polynesia (Polynesians), 21–34, 37–38
   Hokule'a voyage, 27, 28, 31–34, 38
Polynesian Voyaging Society, 28, 30
Popular Mechanics, 67
population movements, 28–29
Porter Hypothesis, 91
Porter, Michael, 91
post-traumatic stress disorder (PTSD), 183, 184
prediction errors, 56, 60–61, 63–65, 237
   dopamine and, 9, 43, 44, 69, 82–83, 234
   free energy principle and, 60–61, 63–64
   reward, 9, 43, 44, 69
   Wundt Curve and, 65, 65–67

predictive coding, 56
predictive processing, 19, 55–67, 171, 209, 219, 230, 238
  bottom-up vs. top-down, 55–56, 70–71
  Dark Room problem, 58–62
  free energy principle, 57–58, 59–61, 63–64, 72
  origins of, 56–57
  sushi roulette, 62–63
preschoolers, 109–10, 116–17, 178–79
Previc, Fred, 233–34
Princeton University, 44, 107, 196
principle of least effort, 208
Prisoner's Dilemma, 91
Psychological Review, 134
Pulitzer, Joseph, 169
purposive behaviorism, 133–35
Putnam, Robert, 171–72, 182
Pytheas, 85

Qesem Cave, 114–15, 122, 128
quantum computing, 6, 150–51, 153–55, 160–62, 164–65, 196, 212
quantum mechanics, 150, 153–54, 161–62
quantum supremacy, 164–65
qubits, 150
Québec City, 226, 232
Quiros, Pedro Fernandez de, 22
"QWERTY effect," 152

Raid the Hammer, 53–55, 70, 72
RAND Corporation, 102, 119
random exploration, 81–83, 86–87, 107, 108, 118, 153
random traveling, 192
Rao, Rajesh, 56
rats, 41, 42, 133–38, 153
Rawls, John, 216
Rayleigh, John William Strutt, Lord, 149
Red Fort, 4
regrets, 88, 202, 237
Reich, David, 49
Reid, Clement, 28
Reid's Paradox, 28–29
relative exploration, 228–29
"replacement crowd," 20–21
research and development (R&D), 162–63, 164. *See also* patents
"restless bandit" problem, 89–90, 122
Reuttinger, Susanna, 93–94

reward and dopamine, 40–41, 42–43, 69
reward prediction errors, 9, 43, 44, 60–61, 69
rhesus monkeys, 46
Rietveld, Erik, 63–64
"Right to Risk in Wilderness, The" (McAvoy and Dustin), 176–77
risk, 177–78, 202
  ambiguity and, 100–102, 108–9, 110
  uncertainty and, 98–99
Rite of Spring, The (Stravinsky), 124–25, 160
Road of Trials, 213
roaming patterns, 191–92
Robbins, Herbert, 88
Robinson, Terry, 42, 43
RoboCup U.S. Open, 67
Rocky Mountains, 2, 9, 104, 105–6, 111
Roepstorff, Andreas, 64
Roest, Patrick, 84–85
Rogers, Kenny, 236
Roggeveen, Jacob, 22
Röjler, Johan, 77–78
Ross, Alex, 124–25
Rother Valley, 190
round-the-world trip of Nellie Bly, 169–70, 173–74, 180–81, 185–87
Rousseau, Jean-Jacques, 160, 189
Routeburn Track, 4
Royal Canadian Geographic Society, 222
RuanRoyal Canadian Mounted Police, 191
Royal Society of Victoria, 137–38, 141
Ruan, Bowen, 62, 63
Rumsfeld, Donald, 99, 100, 103
Runner's World, 5
running, 5, 6, 8, 11, 53–55, 70, 192, 203–4, 206–8, 212, 215–17
Rutherford, Ernest, 149
Ruvalcaba, José, 36–37, 44, 46, 72
Ruvalcaba, Timoteo, 36–37, 44, 46

Sacks, Oliver, 42, 183
Sagan, Carl, 19
Sahtú, 104
St. Lawrence River, 97
St. Louis Post-Dispatch, 169
Sand County Almanac, A (Leopold), 244–45
"satisficing," 90–91
Scandinavian emigrants, 48
Schiller, Friedrich, 189
schizophrenia, 128, 183, 184

Schmidhuber, Jürgen, 159–60
Scholastic Testing Services (SAT), 198
Scholz, Jackson, 206
Schrödinger, Erwin, 154
Schultz, Wolfram, 42–43
Schulz, Eric, 86–88
Schulz, Laura, 194–95
Schwab, Keith, 161–62
Scientific American, 243
screen time, 200–201
Secretary Problem, 92, 93
sedentary lifestyle, 120, 177–78, 200
seeds, 18, 28–29
Seiler, Stephen, 85, 95
Sekani First Nation, 111
Selwyn Range, 144
Sepah, Cameron, 234
September 11 attacks (2001), 99
Shackleton, Ernest, 106
Sharp, Andrew, 21–22, 24–28, 34
Sheehy, Carrie, 188
Sheehy, Cody, 188–89, 192–93, 200, 201, 231
Shoalts, Adam, 221–22, 223–25, 228–29, 230
Shor, Peter, 150–51, 161
Shpayer, Ella Assaf, 114–15, 118–19
Silk Road, 151
Simon, Herbert, 7, 90
Skinner, B. F., 133
Skywalker Ranch, 206
Sloan, Andy, 217
slope-building, 64–65, 123–24, 126, 189–90, 200, 201, 219–20, 223
slope-chasing, 64–65, 68, 72, 123–24, 126–27, 159, 189–90, 219–20, 223
Small, Willard, 132–33, 138
Smith, Adam, 207–8
Smithsonian Institution, 164
Snake River, 245
social bonding and oxytocin, 212–13
social media, 177–78, 211, 223, 234
social network theory, 44–45
Society for the History of Discoveries, 224
Society Islands, 24
somatosensory thalamus, 134–35
Sony AIBOs, 67–69
soulslike games, 218–19
"soups versus sparks," 41
South Pole, 66, 187, 233
space exploration, 19, 52
space race, 163–64

Spanish Pyrenees, 241, 242–44
sparrows, 47
spatial memory, 70
spatial metaphors, 152
speed skating, Van der Poel's story, 75–76, 77–78, 83–85, 94–95
Stanford University, 7
Star Wars (movie), 205–6
Steinbruck, Johann, 97
stimulus-response navigation, 144–45, 148, 182, 183–84
Stone Age, 34
stone tools, 114–15, 118–19
Stony Desert, 141–42
Stravinsky, Igor, 124–25, 160
stress, 69, 238, 239
Strzelecki Creek, 147–48
Stuart Highway, 147
Stuart, John McDouall, 137–38, 140–41, 147, 155, 156
Sturt, Charles, 139, 140–42
suffering, 71
Suits, Bernard, 194–95, 229–30, 233
Superior, Lake, 232
Super Mario 64 (video game), 184
sushi roulette, 62–63
Susquehanna River, 226
Susquehanna Valley, 227–28
Swanson, James, 38–39
Swift, Taylor, 177
Swinburne, Eddie, 203
Sykes, Rebecca Wragg, 18
synapsis, 120

Tahiti, 25–26, 33–34
Taj Mahal, 4
Tasmania, 4
"teasing effect," 63
technological revolution, 122
Tel Aviv University, 114
telegraph, 137–38, 147
tendinitis, 202
Tenzing, Jamling, 175
Terrae Incognitae, 224
Theobald, Johnny, 204
Theory of Fun for Game Design, A (Koster), 218
thigmotaxis, 136–37, 138–39, 144
37 percent rule, 92–94
Thomas Cook & Son, 174–75, 180–81, 186–87
Thomas, George, 190–91
Thomson, J. J., 149, 150

Ticuna, 40
Tikopia, 24
time horizon, 121–23, 127–29, 236–37
　childhood exploration, 115–19, 122–23, 126
　extreme ambiguity aversion, 127–29
　Horizon Task, 108–9, 119–20, 127, 179
time perception, 54
Toanché, 232, 235
Tolkien, J. R. R., 128
Tolman, Edward, 133–35, 148, 152
Tonkin Gulf incident, 102–3
toolmaking, 114–15, 118–19
top-down processing, 55–56, 71
Torrance, Paul, 198–99
Torrance Tests of Creative Thinking, 198–99
Toucan Sam, 159
Trail Runner, 94
trail running, 192
Train, George Francis, 169
Traveling Salesman Problem, 91
treasure hunts, 153
Treatise on Probability, A (Keynes), 100–101
tree seeds, 18
Trollhätten, 75
Tropic of Capricorn, 142
Truffaut, François, 214
Tuamotu Islands, 33
tunnel vision, 213
Tupaia, 25
Turing, Alan, 154
Turner, George, 169–70
Turner Joy, USS, 102
Twenge, Jean, 115–16, 122
20th Century Fox, 121
twin studies, 38, 45–46, 184
two-armed bandit, 79, 107–8

Uber, 132
uncertainty, 2, 61–65, 78, 98–103
　ambiguity aversion. *See* ambiguity aversion
　"feeling of grip," 63–64
　food ordering, 80, 86–88
　risk and, 98–99, 100, 108–9
　sushi roulette, 62–63
　Timothy Wilson room experiment, 61–62
　unknown unknowns, 99–100, 103

uncertainty bonuses, 8, 9, 82, 87, 88, 98, 103, 107, 109–10, 244
uncertainty-directed exploration, 81–83, 86–87, 107, 108, 118, 153
uncertainty sweet spot, 65, 67, 72, 175, 237
uniformity, 118
Unilever, 80
University College Dublin, 49
University College London, 57, 134–35
University of Albany, 45
University of Alberta, 191
University of Arizona, 179
University of California, Berkeley, 116, 132, 133, 134, 209
University of California, Davis, 89
University of California, Irvine, 239
University of California, Los Angeles, 93
University of California, Santa Barbara, 27–28
University of Cambridge, 79–89, 149
University of Chicago, 156
University of Georgia, 199–200
University of Girona, 18
University of Guadalajara, 44
University of Hawaii, 21, 28
University of London, 26
University of Michigan, 42
University of Minnesota, 198
University of Oxford, 34, 150–51, 153–54
University of Pennsylvania, 203
University of Rochester, 66
University of St. Andrews, 138
University of Strasbourg, 162
University of Texas, 233
University of Toronto, 59, 208, 223
University of Tübingen, 117
University of Virginia, 61–62
University of Washington, 50
University of Waterloo, 109, 178
unknown unknowns, 99–100, 103
Upper Confidence Bound algorithm, 88, 237
Uppsala University, 218
"Usefulness of Useless Knowledge, The" (Flexner), 162–63
"use it or lose it," 120
Usherwood, Stephen, 174–75

Van der Poel, Nils, 75–76, 77–78, 83–85, 94–95

Van Gogh Museum (Amsterdam), 156–59
Verne, Jules, 169, 174, 180
Victorian Exploring Expedition, 130–31, 139–44, *143*
video games, 60, 66, 68, 183–84, 218–20, 223, 234
Vietnam War, 100, 102–3, 107
view enhancing, 192
Vikings, 11–13
Vinland, 11–12
virtual reality, 138, 145, 170–71, 222
von Neumann, John, 162
VO_2 max, 54

Waddington, Emma, 70
Waldeyer, Wilhelm von, 162–63
Wallace Line, 21
Wall Street Journal, 94
Wang, Dashun, 156–59, 237
War and Peace (Tolstoy), 7
Washington Post, 216
water crossings, 34
  of Polynesians, 21–38
waterfall descents, 99, 221–22, 223–24
Watergate, 100
Watson, John, 133
wayfinding, 32, 138, 153
Wealth of Nations, The (Smith), 207–8
Western Brook Pond, 2–3
Whittle, Peter, 79–89
Wikipedia, 155

Wilderness Act of 1964, 227
Wilderness and the American Mind (Nash), 227–28
wilderness zones, no-rescue, 176–77
Williams, John, 24
Willmore Wilderness Park, 3–4
Wills, William, 130–31, 132, 137–38, 140, 141–44, *143*, 146–48, 156
Wilson, Robert, 8, 107–9, 116, 127–29
  extreme ambiguity aversion, 127–29
  Horizon Task, 108–9, 119–20, 127, 179–80
Wilson, Timothy, 61–62
Wind River Range, 176–77
Winston, Patrick, 56
wobbling plate, 197–98
wolves, 18, 29
Wordsworth, William, 149
World War II, 80, 163, 196
Wu, Charley, 117–18, 120–21, 126–27, 151, 153, 236
Wundt Curve, *65,* 65–67, 72, 124, 175, 176, 185
Wundt, Wilhelm, 66

Yale University, 39, 101, 158
Yukon, 4, 245

zaffs, 117
Zak, Paul, 212–13
zebra finches, 83

ABOUT
MARINER BOOKS

MARINER BOOKS traces its beginnings to 1832 when William Ticknor cofounded the Old Corner Bookstore in Boston, from which he would run the legendary firm Ticknor and Fields, publisher of Ralph Waldo Emerson, Harriet Beecher Stowe, Nathaniel Hawthorne, and Henry David Thoreau. Following Ticknor's death, Henry Oscar Houghton acquired Ticknor and Fields and, in 1880, formed Houghton Mifflin, which later merged with venerable Harcourt Publishing to form Houghton Mifflin Harcourt. HarperCollins purchased HMH's trade publishing business in 2021 and reestablished their storied lists and editorial team under the name Mariner Books.

Uniting the legacies of Houghton Mifflin, Harcourt Brace, and Ticknor and Fields, Mariner Books continues one of the great traditions in American bookselling. Our imprints have introduced an incomparable roster of enduring classics, including Hawthorne's *The Scarlet Letter*, Thoreau's *Walden*, Willa Cather's *O Pioneers!*, Virginia Woolf's *To the Lighthouse*, W.E.B. Du Bois's *Black Reconstruction*, J.R.R. Tolkien's *The Lord of the Rings*, Carson McCullers's *The Heart Is a Lonely Hunter*, Ann Petry's *The Narrows*, George Orwell's *Animal Farm* and *Nineteen Eighty-Four*, Rachel Carson's *Silent Spring*, Margaret Walker's *Jubilee*, Italo Calvino's *Invisible Cities*, Alice Walker's *The Color Purple*, Margaret Atwood's *The Handmaid's Tale*, Tim O'Brien's *The Things They Carried*, Philip Roth's *The Plot Against America*, Jhumpa Lahiri's *Interpreter of Maladies*, and many others. Today Mariner Books remains proudly committed to the craft of fine publishing established nearly two centuries ago at the Old Corner Bookstore.